About the Editor

BILL FAWCETT is the author and editor of more than a dozen books, including *You Did What?* and *How to Lose a Battle*. He is also the author and editor of three historical mystery series and two oral histories of the U.S. Navy SEALS. He lives in Illinois.

D0348821

It L⌢ked Good on Paper

ALSO BY BILL FAWCETT

It Lked
Good on Paper

Bizarre Inventions, Design Disasters,
and Engineering Follies

EDITED BY

BILL FAWCETT

HARPER

NEW YORK • LONDON • TORONTO • SYDNEY

HARPER

HarperCollins books may be purchased for educational, business, or sales promotional use. For information please write: Special Markets Department, HarperCollins Publishers, 10 East 53rd Street, New York, NY 10022.

FIRST EDITION

Designed by Joy O'Meara

Library of Congress Cataloging-in-Publication Data is available upon request.

ISBN 978-0-06-135843-2

09 10 11 12 13   OV/RRD   10 9 8 7 6 5 4 3 2 1

Contents

Introduction

Welcome to the

Wonderful World of Failure

Winston Churchill once said that "success was going from failure to failure without losing your enthusiasm." Gathered in this book is a collection of flawed plans, half-baked ideas, and downright ridiculous machines that, with the best and most optimistic intentions, men have constructed throughout history. Some failed spectacularly, others fizzled after great expense, one crashed on Mars and a few of these amazing broken ideas are still in use today. While this is a book about ideas and things men make that have failed, it is not a technical manual. Rather it is a fun collection that shows how an otherwise brilliant designer, scientist, architect, or doctor can often spend years or millions of dollars creating something that a few small bits of common sense would have prevented. This book begins looking at some of history's mad plans and ends with an area that has seen more than probably any other really stupid designs, the military. In between we look at autos, medicine, some really bad government plans, planes that didn't fly, and ships that sank.

Feel free as you read to wonder how anyone as brilliant as many of the men who created these masterpieces of failure could have gone so far down the wrong path. These are not the creations of fools, but of men like Edison, top auto designers, the man who created the first airplane, rocket and nuclear scientists, and many others whose names you will recognize as revered thinkers. And you will be able to picture them, like any lesser inventors throughout history, just shrugging their shoulders with a wan smile as their bright idea fails and murmuring that final excuse, "It looked good on paper." And keep in mind, despite all of these failed plans progress has continued to march on. Probably chuckling knowingly as it passes them by.

"Success consists of getting up just one more time than you fall."

—Newt Gingrich, former Speaker of the U. S. House of Representatives

Past Imperfect

We tend to think of crazy designs and badly engineered structures as modern phenomena, but they have been around as long as man has strived to create the new and different. Somewhere still buried in a cave has to be the prototype of a square wheel.

"Pride and conceit were the original sins of man."

—Alain Rene Lesage

The Great Stele of Aksum

E. J. Neiburger

The rise and fall of many civilizations can often be traced to well-planned projects that looked pretty good in theory, but failed miserably upon execution. The story of the Great Stele of Aksum is a wonderful example.

The biblical Queen of Sheba (of Solomon and Sheba fame) built her fifty-two-room palace and ruled her great empire from the city of Aksum, in what is now northern Ethiopia. Two millennia after her death, another great civilization and world power arose from the skills and savvy of the Aksumites who created the ancient Kingdom of Aksum. This world empire extended from Ethiopia and Sudan over the Red Sea and into southern Arabia. It matched, and in some cases exceeded, the size and accomplishments of ancient Egypt, Rome, Greece, and Persia.

The kingdom began in the highlands of northern Ethiopia, the commercial crossroads for trade from Europe and Egypt into Africa. It was also the main trading terminal (seaport) for commerce from Asia, Arabia, and India into east and central Africa.

The Sudanese desert and Rift Valley funneled most of Africa's trade through the rich merchants and rulers of Aksum. The land was very productive, well watered with carefully tended dams, wells, and reservoirs.

The kings of Aksum grew rich on the grain, skins, animals, frankincense, spices, iron, gold, and especially ivory supplied by Ethiopia's forty thousand elephants, and all this merchandise headed north, out of Africa, into Egypt and on to Arabia, India, Syria, and Europe. They grew filthy rich on metals (gold, silver, and iron), wine, olive oil, and other riches from the northern and eastern lands, which passed through their territory on the way to Africa. The heyday of Aksum was from AD 300 to AD 600. With all this wealth, Aksumite society became well educated, organized, and technically and artistically advanced. They were "hot stuff" and wanted to flaunt it.

In Africa, there is a five-thousand-year-old tradition that rich people build big monuments, among the biggest, most ostentatious of all monuments, were obelisks. In Ethiopia, these are termed steles. Now these steles were not the usual obelisk structures made in Egypt, or later such as in the U.S. (Washington Monument). They were not made out of bricks and mortar. No, these were super steles made out of the hardest black granite, each made in one gigantic piece—no cheap bricks. It was all or nothing. Not only were the steles made out of hard rock that was difficult to work, and of a single piece (most difficult and expensive to fabricate), but they also had to be expertly carved and designed. To top off the display of wealth, an enormous stele would require thousands of men to build, move, and erect it, thus attesting to the wealth and power of the kings who created the project—or at least, paid the bills for it.

The ancient Egyptians erected twenty-two known obelisks

(steles) of which thirteen now stand in Italy. The greatest standing Egyptian obelisk was the 32.16-meter (98 feet) stele erected by Pharaoh Tuthmosis III in Karnak, Egypt. The 1200-ton, 41.7-meter-long Aswan obelisk did not count because it cracked when it was only three quarters completed (during manufacture), was abandoned, and never got out of the quarry. The Aksumite kings planned to outdo the Egyptians with a bigger and greater stele. Better than the 1500 BC Tuthmosis obelisk.

Not only was this a point of flaunting one's wealth and hailing the greatness of the Aksumite kings, but the stele took on a religious role as well. The Aksumites believed the stele would actively protect the kingdom by piercing the sky and dispersing the negative forces that create storms and empower evil gods to harm the people. In a pagan society, like early Aksum, such concerns were real magic and had to be dealt with. Also, around AD 300 a new religion, Christianity, was staking a foothold in the kingdom and threatening to destabilize the historic order. A firm statement that the pagan gods and kings were in power had to be made. The largest, highest, and most expensive stele in the world seemed to offer a logical and magnificent solution.

So, around AD 330, the Great Stele of Aksum was carved out of granite to join the other smaller seventy-four Aksum steles already erected. This great stele was a billboard designed to announce the power and authority of the ruling dynasty, blessed by the gods and heaven. It was intricately carved, possessing twelve faux doors and windows and polished to a velvetlike smoothness. It weighed 517 tons and stood 33.3 meters (100 feet) high—technically, taller than the largest Egyptian obelisk. In those days, bragging rights were everything. It is said that the gods (or God) carried it from the quarry to the steles field four kilometers away. As it was erected, it stood straight and tall. It was magnificent. The king, the people, and even the pagan gods applauded—for a short while.

There is no record of how long it stood, but the modern estimate, made by scholars and engineers, is a few hours to even a few days. Not long. The Great Stele, inadequately supported by a foundation much too small for its weight, began to list—and then fell flat on its face. The great monument to the kings and their religion fell on the rocky ground and broke into six massive pieces. It must have been devastating beyond measure, because even the kings could not muster enough support and manpower to cart away the rubble that still lies in the middle of the Stele Field in central Aksum.

Talk about a big flop? This was the most abrupt in the ancient world. The message was clear. The kings were weak. The gods were weaker. People were reminded of this whenever they passed the fallen monument. No more steles were built after that time. Paganism quickly declined. And Christianity (with a little Judaism) flourished unrestricted until the coming of Islam in AD 700. The Great Stele looked magnificent on the drawing board, but there were a few problems in its execution. The result sent a completely opposite message to the world from the one intended—at prodigious cost.

The Pipes of Rome

Chris Powers

Ancient Rome. Two simple words, and they evoke images of implacable legions marching over hairy half-naked barbarians, an empire that spanned the Mediterranean world, crazy emperors, straight roads, bathhouses, central heating, plumbing. Hold that thought.

The Romans were the master engineers of the ancient world. They planned their cities, built huge public works with the aid of concrete, an invention that is still shaping the modern world, and constructed temples and aqueducts that are still standing today.

Eleven aqueducts brought water to Rome through a gravity-driven system of waterways and filtration sumps that fed the public fountains and baths as well as private houses. A network of pipes lay under the city, supplying the private houses of the rich and middle class with all the water they needed for drinking, washing, and flushing—yes, flushing—toilets. Pipes and sewers took it all away again, mostly into the nearest river. That pattern, to a lesser degree, was repeated throughout the empire from Syria to Britain.

At first these pipes were lengths of hollow tree trunks, their tapered shape ideal to slot narrow into wide to form continuous watertight lengths. Then the potters saw a niche market they could exploit, and produced lengths of earthenware pipes that were linked the same way as the tree-pipes. The advantages of these were obvious. They did not rot, and could be made comparatively quickly and easily in different lengths and conformation to cope with corners. The disadvantage was that heavy traffic and very cold weather caused breakages—but the damaged sections could be swiftly repaired. In theory. So the water companies that are today digging up roads of London and other cities to repair broken pipes have a long and noble history.

The next breakthrough solved all problems. In theory. Lead. In conjunction with silver, zinc, and copper, it was mined all over the empire. There were slaves aplenty to do the very dangerous mining, and with a melting temperature of 327.46°C (621.43°F) it was easily extracted from the ore and refined. It was shipped from the mines already formed into large ingots, called pigs for their distinctive shape.

Pouring lead into molds for all kinds of household and wine-making utensils was commonplace, and the plumbing industry simply added another way of exploiting its versatility. So before long, crack-proof lead pipes delivered rain water to cisterns, which were frequently lead lined. Lead pipes took water into the baths, houses, and tavernas.

The use of lead, therefore, was uniform and widespread. As lead acetate it was even used as a sweetener in food recipes and wines. All of this in spite of the fact that even back then it was recognized that lead was a toxic metal.

Lead poisoning was recognized by civilizations in the fertile crescent way back in 2000 BC. It was certainly well known in Rome. Julius Caesar's engineer, Vitruvius, reported on it, and Aulus Cornelius Celsus included it in his list of poisons along with hemlock. In large doses, it is lethal and they knew it. On a lesser scale the damage it can cause is pretty extensive—aggressive and irrational behavior, gout, kidney damage, anemia, seizures, severe and permanent learning disabilities, and male sterility to name a few.

Which brings us right back to the crazy emperors. It has to be said that Rome seemed to have had more than its fair share of lunatics in charge of the asylum, but the symptoms went right through Roman society from top to bottom. But the population grew and there were sesterces to be made which resulted in new construction projects that inevitably meant new plumbing, which meant more lead pipes and cisterns.

Given that the toxic nature of lead was known, yet its usage was not banned or curtailed—an insanity in its own right—it is highly likely that it wasn't the descendants of the hordes of hairy and half-naked barbarians who brought down the empire. It was the Romans' own inability to turn away from expediency and the easy option that damaged them from within. Then all it needed was a shove in the right place, and Rome fell, pipes and all.

> "The best laid schemes o' mice an' men gang oft agley."
>
> —Robert Burns, 1759–1796

The Great Wall of China

Bill Fawcett

It seems a little strange to include something as enduring as the Great Wall of China in this book. Yet in a very real way the Great Wall was an abysmal failure. In ways other than its primary purpose the wall was a success, not to mention an amazing tourist attraction. But the Great Wall of China never was able to accomplish what it was designed to do, keep the steppe barbarians out of China.

The wall was begun in 214 BC by the first emperor of a united China, Qin Shi Huangdi. It was the most ambitious and costly construction ever attempted up to that time. The sheer scope of the job meant that long sections of the original wall were created by basically connecting the earthen defense walls that already existed into a continuous line. To do this the emperor created a forced labor pool of about three hundred thousand workers. So initially, and for a very long time, the Great Wall was constructed of very long and carefully maintained piles of dirt with the occasional stone fortification scattered along it. It was not until the days of the Ming Dynasty, two millennia later, that the imposing stone structure virtually everyone has seen pictures of was built.

Now the first emperor of China was concerned with creating a permanent nation state with a sense of identity. He began his rule by conquering six other Chinese states, some centuries old. National projects and imposing buildings did help give the Chinese a new sense of being, well, Chinese. What they did not do was to give new Chinese any loyalty to their emperor. In 210 BC a general facing punishment chose to revolt. Many people resented the costs in lives and money squandered by Qin Shi Huangdi, who had just died, and joined the revolt. The Qin Dynasty quickly fell and the Han Dynasty took power. Over the centuries, China would again split into competing states on several occasions, but the sense of identity as Chinese remained.

So in this way the Great Wall, and Qin Shi Huangdi's many other construction projects, were a success. But as a way to ensure China's security against the nomadic tribes to the North, history was to show it a failure. A big, long, very expensive failure. The first problem was that even the dirt-packed Great Wall and its forts were expensive to maintain. This meant often they simply were not kept in repair or well manned. The second problem was that in 1234 the Mongols invaded. The Great Wall was designed to keep out the steppe tribes, but not when they appeared as a well-organized army in great numbers. The Mongols barely paused at the Wall before invading China. Eventually all of China, including the Great Wall, was controlled by the Mongols.

In 1368 the Chinese, under the rule of the Ming Dynasty, completed the wall that we know today. Three hundred years later another army of steppe barbarians, the Manchus, again waltzed right through the wall and took over. They, the descendants of the steppe barbarians, remained in control of China until 1911, when the Nationalist Chinese state emerged.

So the Great Wall of China was an engineering and architectural wonder, but it never really was much use in stopping the barbarians from taking over the country.

The Tower of Pisa

Chris Power

The date is August 9, 1173. The place is Pisa, in Tuscany, Italy. By the twelfth century, the city was the heart of a rich maritime republic with a history that stretched back over a thousand years. It grew up around the joining of two rivers, the Arno and the now-vanished Auser and over the centuries its power and importance grew.

Inevitably, Pisa had rivals. Their main competitor was Florence, with Genoa a close second, and to say diplomatic relations were strained would be an understatement. They fought wars. They also indulged in bouts of one-upmanship on a grand scale.

In 1172 a rich widow named Berta di Bernardo died, and legend claims that in her will she left sixty coins of unspecified value to pay towards the construction of a campanile for Saint Mary's Cathedral in the Piazza dei Miracoli. St. Mary's already had a baptistery to match the splendid cathedral and it seemed like a really good idea to complete the set with a bell tower. In fact, the Pisans went the whole hog and decided to make it a belvedere, a tower you could walk up the outside of as well, where lucky citizens could watch the processions and celebrations that regularly took place in the piazza in front of the cathedral.

This tower would be a masterpiece of gleaming marble, raised for the glory of God and proclaiming the superiority of Pisa and its people to all. Especially to Florence.

Plans were made, the site in the Piazza dei Miracoli was chosen, and on that August day, work began.

First the foundations were dug. These consisted of a circular ditch approximately three meters deep. This was meant to be the base for a structure that would weigh more than 14,000 metric tons and stand 55 meters high. Maybe it would have worked— if the substrates had not been river clay and sand. Remember those two rivers?

Even before the first three levels were finished, the problems that would trouble the tower had begun. By 1178, as the weight distorted the foundations, it began to list to one side.

What the architect thought of this is not known. Indeed, no one knows for certain who he was. It may have been a case of whoever it was not wanting to have his name associated with the debacle, because there is no doubt that the building contractors had made a hash of it from the word go. Even in the twelfth century, most knew about bedrock and clay with what the difference could or couldn't support in the way of structures.

So there they were, red faces all round, the Florentines laughing themselves sick, and the Pisans' beloved status symbol about to fall over on them. Since they couldn't stop the tower from leaning, upper floors were actually made with one wall higher than the other to give the illusion of the tower being straighter. Indeed, 1178 was a bad year, because as well as laughing, the

Florentines also declared war on Pisa. All construction on the tower stopped, but not for long. Peace broke out for a while, but in 1185 Pisa and Florence clashed once more and work stopped again. This time the delay was longer. Much longer. Work didn't recommence until 1272. Then there was another halt and a major sea battle with Genoa in 1284—which Pisa lost.

In 1319 the tower was finished at last, even though the full ring of bells weren't installed until 1350, and that was that. Apart from the increasing list.

For centuries architects and builders tinkered with the tower, all to no avail. Once again common sense was ignored in favor of public aggrandizement when another bell, the largest yet, was added. It weighed in at 3.5 tons and increased the pressure on the beleaguered foundations.

By 1817, the tower had an incline in the region of five degrees. Some twenty years later a particularly bright spark named Alessandro della Gherardesca decided it would be a very clever idea to display the foundations to the general public. So he had a trench dug around the base of the tower. Remember the river? His trench cut below the water table and it flooded. To the extreme fascination of the aforementioned general public, Pisa's tower attained an additional meter of list in the space of a few days.

Even so, it was not for another hundred years before serious attempts were made to prevent the tower falling. In 1934, holes were drilled in the foundations and a kind of grout was forced into them. This had a reverse effect and the result was another increase in the tilt. Furthermore, 1966 and 1985 saw other attempts, with a similar outcome. The tower's incline was now so severe it seemed that its ultimate collapse was imminent. In 1990 the tower was officially closed to the public on safety grounds.

But no one was prepared to give up on the Leaning Tower of Pisa. In the latter end of the twentieth century, a more sophisticated technology was brought to bear on the problem. First a corset of steel was fixed around the base of the tower, then a series of concrete foundations and counterweights were set up to pull back some of the tilt. After a major miscalculation that nearly finished the whole project in a cloud of dust, the experts hit on the idea of undermining the north side of the tower and taking away the soil. Rather than building up the side that it was tilting to, they would simply encourage the tower to tilt back the other way. This actually worked. The incline was substantially reduced and, more important, stabilized. On December 15, 2001, the Leaning Tower of Pisa was opened to the public once again.

So there, at a dignified if slightly drunken angle, it stands. The Campanile of St. Mary's Cathedral in the Piazza dei Miracoli, first commissioned by Berta di Bernardo in 1172. Against all logic it has survived trials and tribulations and the idiocy of experts, and is justifiably a major tourist magnet.

Yes, it looked good on paper, but the odds were stacked against it. Yet at the final reckoning the grand tower not only survived its initial design flaw, but also three disastrous fixes that made matters worse.

"In Georgia, the legend says / That you must close your windows / At night to keep it out of the house."

—James Dickey, from "Kudzu"

Overwhelming Success

Bill Fawcett

For the first one hundred years there was no kudzu growing anywhere in the United States. This voracious plant first appeared at the 1876 Centennial Exhibition, which was held in Philadelphia. Many countries build large and extravagant exhibitions to honor the young nation's centennial. Among these was Japan, whose hall was filled with beautiful and exotic plants. Among them was kudzu, which has large leaves and sweet-smelling blossoms. The plant was an instant hit among the gardeners who toured the Japanese Pavilion. By 1920 you could buy kudzu plants through the mail or at nurseries all over the South.

Today Kudzu covers *seven million* acres across the southern United States. What is worse, when it covers, it covers everything, even climbing and strangling the tallest trees for light. This vine is now considered to be a major menace to native plant life. Where kudzu rules, no other plant survives. But this didn't just happen; we did it to ourselves.

During the Depression, one of the ways in which the

members of the Civilian Conservation Corps worked was to plant kudzu to prevent erosion. After all, the kudzu plants grew quickly and covered the areas with roots. The CCC even paid an incentive of eight dollars per acre planted with kudzu. By 1946 almost three million acres had been "conserved" with kudzu planted by the government and the vine could be found all over the South. It was ideal for this use, unless you ever wanted to use that land, and later any adjacent land, and a bit later most of the nearby land for anything else. For a while kudzu was considered a miracle plant. Goats ate it, some medicines might be derived from it, the vines made great baskets, and the leaves were even marginally edible. There was a Kudzu Club with almost twenty thousand registered members. But soon kudzu had begun creeping across whole forests and fields, blocking out the sun from every competing plant and taking over anywhere it grew. And without any of the insects or other parts of the ecology, including hard frosts which limit the vine in its native Japan and keep it from spreading too far north in the United States, nothing slows its spread. Kudzu is one of the fastest-growing vines commonly found anywhere. It can spread at the amazing rate of a foot a day, and sixty feet a year is not unusual. And where it spreads, every other plant is gone, cut off from the sun and dead.

By 1953 even the government stopped advocating any further planting of kudzu. Research gradually changed from finding new uses for the rapidly spreading plant to finding ways to kill it or slow its growth. This has proven surprisingly difficult. Most herbicides don't touch it and a few even make it grow faster. With the exception of flooding each field of kudzu with hundreds of goats who will literally eat it to the ground, nothing has been found that can harm the hardy plant. In 1972 the

Department of Agriculture formally declared kudzu a weed.

The once subsidized and praised kudzu is now described using terms like "annoying," "intruding," and "menace." Using the Japanese vine for erosion control looked good on paper, but too much of anything is just . . . too much. And they planted it on purpose.

The Sword Pistol

Bill Fawcett

Around 1800, the pistol was a popular weapon, carried by police, merchants, and gentlemen. One problem with the flintlock pistols of the day was that they were finicky. They had a tendency to misfire, often. Since they were so unreliable, a unique hybrid weapon was briefly popular around the same time. This was the sword pistol. The combination was created by attaching a blade, often more than a foot long, to the side of flintlock pistol. This blade was hinged to fold back against the pistol until needed and then was clicked forward into place. The theory was that the user now had a small sword, often nearly three feet long, to use after firing or if there was not time to fire the flintlock pistol.

It seems likely that more than a few of those who carried this combination weapon discovered its shortcomings the hard way. If you did not fire before using the blade, the action of deploying the blade was likely to knock the powder from the primer pan and ensure the pistol part of the weapon was useless. But if you fired first there were more problems. These included weight, nearly three pounds that had to be balanced and aimed with one hand. Combined with the windage of the barrel in a pistol of that era, this nearly guaranteed a miss. The next problem was that, after firing, flipping the blade forward and locking it into place took time. Time you would not have in a close quarters melee. The next problem was that while you could stab with a pistol sword, the edge was only inches long, limiting its use as a sword. Then there was the heavy pistol part now doubling as the handle of a sword, but the pistol is gripped in a way that hardly makes for robust use of the blade or point. This same configuration made the sword pistol almost useless for blocking the attacks of your opponent, particularly if he had a simple knife or other nimbler weapon. Like all weapon systems that try to be too many things at once, the sword pistol was not effective either as a sword or a pistol. Perhaps the most positive thing that can be said about this weapon is that it was superior to the combination pistol, dagger, and brass knuckles (finger holes and all) known as the Apache pistol that was used in America by street gangs in the 1880s.

The True Saga of the Pony Express

Douglas Niles and Donald Niles, Sr.

One of the cherished legends of the American West is the story of the Pony Express: brave horsemen who rode at a fast clip from Missouri to California, carrying mail to the rapidly growing population in that newly established Pacific coast state, stopping only to change horses, braving the weather, escaping hostile Indians and outlaws. It is a story based on truth, but it is an episode of history that actually lasted for only a few short months.

Of course, that might be considered particularly appropriate for this aspect of California history, since that state, especially in its early years, was a place where things happened very quickly indeed. In 1848, California was a lightly populated territory of the United States, having been claimed from Mexico as part of the settlement of the Mexican War. When gold was discovered—a discovery noted by President James Polk in his state of the union report in December of that year—a gold rush was unleashed that swelled the population and infrastructure in the San Francisco Bay area to the extent that California was granted statehood in 1850.

Almost immediately, the problem of communicating with this distant state took on an air of urgency. In 1848, it took some six months for a message from Washington to reach San Francisco. There were no good routes: the Oregon Trail and other land routes were plagued by hostile natives, grueling mountain and weather obstacles—it was thought that the trip was utterly impossible during the winter—and vast deserts. A message carried by ship needed to be carried around the stormy tip of South America at Tierra del Fuego, which was never an easy task. In fact, it was not unheard of for a sailing captain to give up the fight against the constant storms, and turn around to take the long way to California, crossing the Atlantic, Indian, and Pacific Oceans. Another alternate route meant sailing into the Caribbean Sea, carrying the message over the mosquito-infested mountains of Panama—with the attendant risks of malaria and yellow fever—and then another long sea voyage from Central America to Northern California.

As the population of California continued to grow, a rail connection to the west coast remained a pipe dream—the existing tracks didn't extend beyond St. Joseph, Missouri, on the great river for which that frontier state had been named. The telegraph was still in its infancy, and for the time being, no one was considering linking the east and west coast with wires and telegraph poles. The written letter remained the standard form of remote communication, and a letter had to be physically carried from the sender to the recipient.

But where others saw obstacles and challenges, several directors of the Central Overland California and Pike's Peak Express Company firm saw only opportunity. Three men, William Russell, Alexander Majors, and William Waddell, were willing to invest some $700,000 to gain the answers to key questions. Was it possible to establish a route to the west coast that could be employed even during the winter? What if a network of stables could be created, extending from Missouri to California? What if bold riders could travel at a gallop, carrying little more than a mail bag and a revolver, trading horses every ten miles or so? How long would it take a letter to travel from the end of the railroad line, at St. Joseph, to a steamer landing on the Sacramento River, from which the mail could be carried by ship down to San Francisco?

The answers were yes, the trip was possible, the network of stations could be established, and in fact the route could be covered in a remarkable ten days.

The Pony Express was established in early 1860. It included a network of more than 150 stations spaced from five to twenty miles apart. About a hundred riders were hired for the good salary of $100 per month. All were male, with some as young as their early teens. The maximum weight allowed for a rider was a slight 125 pounds. The "help wanted" ad posted in California sought riders who were "skinny, wiry fellows" willing to face death on a daily basis, and even suggested that orphans were preferred. The mail would be carried at a gallop, with the horse changed about every ten miles, and each rider averaging about a hundred miles per shift. (The longest single leg was a buttocks-numbing 370 miles ridden by the legendary Pony Bob Haslan!)

The total route covered more than 1800 miles, and included portions of the Oregon Trail, the Mormon Trail, and the California Trail. Beginning at St. Joseph, in northwestern Missouri, the route traversed the plains to Fort Kearney, Nebraska. From there it followed the route of the Platte River (much as I-80 and I-76 do today) to Julesburg, Colorado. From Julesburg it veered north into Wyoming, past the important fort at Laramie, tracing the course of the Sweetwater River to Fort Caspar, then turning through South Pass to wind its way to Salt Lake City. Moving through the arid wastes of the Utah and Nevada desert country, the trail crossed the lofty Sierra Nevada Mountains near Lake Tahoe, descending to Sacramento, where the mail could be put aboard a steamship for the final leg to San Francisco.

The horses were not actually ponies, but instead numbered some 400 pintos, mustangs, Morgans, and even a few thorough-breds. In general the steeds were small and hardy, capable of traveling through rugged country, possessing the endurance to travel at a gallop for the ten miles or so required for each leg of the ride. Many legendary western characters signed on to ride for the Pony Express, the most famous being William "Buffalo Bill" Cody. They carried the mail in a saddlebag known as the *mochila*, and postage was charged in the amount of $5 per ounce initially, though the cost was soon lowered to $1 per half ounce in an effort to entice more customers.

The fastest run of the Pony Express carried a copy of Abraham Lincoln's inaugural address across the country in less than eight days. The exact numbers of express runs is unknown, though it is commonly understood that only one rider, and one *mochila*, were lost during the entire run of the Pony Express. The experiment determined that the route to the west could be employed during all seasons, and helped to map out the path that would later be used for the transcontinental railroad.

Unfortunately for Russell, Majors, and Waddell, their company failed to win the government contract to carry the mail on a long-term basis. The telegraph was coming into its own, and by 1861 the network of poles and wires was quickly being extended across the continent, and would soon allow for nearly instant coast-to-coast communication. By November 1861, the firm had run out of money, and the Pony Express was disbanded as obsolete after a brief, but glorious, run of some 18 months. The closing was announced in October of that year, not coincidentally coming two days after the Transcontinental Telegraph reached Salt Lake City.

The three founders lost several hundred thousand dollars, and were forced to sell out to Ben Holladay, who was operating the Butterfield Stagecoach Line. (After the Civil War concluded, in 1865, Holladay was able to sell his own investment to Wells Fargo for more than 1.5 million dollars.)

In a business sense, the Pony Express was a failure, but it was a failure in grand style. Even today, the image of those solitary riders and their doughty little horses remains a fixture in Americans' deep connection to their western heritage.

"It's my invention, dammit!"

—Thomas Alva Edison

Thomas Edison's Insistence on the Use of DC Power

Douglas Niles and Donald Niles, Sr.

Thomas Edison is well known as an inventor and a practical scientist. He personifies the idea of Yankee ingenuity, working in a lab that he owned, uncovering many secrets—especially related to electricity—and designing machines, equipment, and technology to put his knowledge to practical use. He obtained more patents for his inventions—better than a thousand of them—than any other person in history.

Working during the latter part of the nineteenth century and the early part of the twentieth, Edison lived during a time of great technological accomplishment. As is the case with so many of America's self-made successes, Edison came from relatively humble beginnings. As a boy he worked on the trains running between Port Huron and Detroit, Michigan. He was a poor student, in part because he was not interested in the rote memorization that was typical of education in the mid 1800s.

Also, he suffered from a hearing impairment, which impeded his progress in school. This partial deafness also motivated many of his inventions.

His first serious job involved working as a telegrapher, beginning in 1863. At that time, telegraph messages were communicated onto a printed piece of paper, with the dashes and dots of Morse Code read by the telegraph operator. As the technology improved, however, the telegraph system increasingly relied upon the operator's listening to the signals as they came over the line. Edison's hearing difficulties made his career choice increasingly difficult, though he worked for some six years as an itinerant operator throughout much of the United States and Canada.

By 1869, however, his ingenuity and frustration combined to put him on the path that would become his life's work. Instead of operating the telegraph, he put his talents toward making the existing equipment work better. His first creation was a duplex telegraph, which allowed two messages to be sent over the same wire simultaneously, and a printer that would render the electronic symbols from the telegraph line into letters and other symbols. These successes caused him to leave the field of telegraphy to devote his life to invention and entrepreneurship.

Moving to New York City, he began to create a dazzling series of inventions. Many of these related to continuing developments in the field of telegraphy, which was dominated in the United States by the Western Union Telegraph Company. Edison was soon perfecting a quadruplex telegraph machine, one that could send up to four messages at the same time. One of Western Union's bitter rivals offered the inventor $100,000 for the device, and—despite the bitter legal wrangling that followed—Edison made the deal and was well on his way to the technical and entrepreneurial successes that would establish his enduring reputation.

Moving to Menlo Park, New Jersey, Edison designed and built a laboratory to his specifications. Here he continued his creative endeavors, leading to, among other things, the phonograph, the electric light bulb, and several key components that would make motion pictures possible. His inventions laid the groundwork for the budding office machine industry, and his understanding of chemistry—coupled with his growing skill with electricity—allowed the first mimeograph machines to take shape. The amazing phonograph, in particular, brought him a whole new wave of publicity. Edison became known as the Wizard of Menlo Park, as his inventions astounded the world and brought him ever increasing fame.

Edison was a skillful negotiator, and arranged many lucrative contracts for himself, but he was a poor money manager. His diverse interests kept him moving from one interesting project to another during an unprecedented age of invention and innovation. His ideas helped lay the groundwork for the telephone, for which Alexander Graham Bell would gain the patent. Edison had a good working relationship with Henry Ford and was an important early investor in the General Electric Corporation.

During this period of history, the generation of electricity was not yet something that was considered on a city-wide or even neighborhood basis. Rather, electrical power was generated usually within, and for, a single facility. But the success of Edison's light bulb, among other things, caused people to think, more and more, of the usefulness of bringing electric power to broad areas.

In September of 1882, the Pearl Street Power station, designed and built under Edison's tutelage, began to operate in

New York City. The station generated electricity at the level of 110 volts, and used direct current (DC) to send the power to the surrounding area. This worked all right for illuminating light bulbs, but several problems with DC power began to manifest themselves to those who sought to employ Edison's power.

For one thing, DC power was difficult and expensive to send over long distances of wire. Additionally, many industrial users of electricity wanted to run large motors that didn't function well, or at all, with the 110-volt power source. As more and more people in more and more places clamored for electrical power, the disadvantages of DC power became increasingly obvious.

George Westinghouse was another inventor and entrepreneur, though not quite as well known—or experienced in the field of electricity—as Edison. Nevertheless, Westinghouse formed his own electric company in 1886 as a direct competitor to Edison's own. The genius behind Westinghouse's work was a Serbian immigrant, Nikola Tesla, who had obtained a number of patents and was working assiduously on the development of alternating current (AC) power. Unlike DC power, AC could be broadcast over long distances with little loss in power. Furthermore, with the use of transformers, AC power could be employed in a wide variety of voltages, allowing the use of many sizes of electric motors as well as the basic illumination allowed by DC power.

Edison's massive Pearl Street station generated a lot of electricity, but none of it was broadcast farther than about a half mile from the installation. If all of New York City was to be powered by DC, dozens, or even hundreds, of large generating stations would have to be built right inside the city. Westinghouse's AC power, on the other hand, could be generated outside the city, economically transported at very high voltages into the metropolis, and then, through the use of transformers, distributed at safe levels to the actual user. The advantages of AC power were clear, and obvious to anyone who understood the technology.

Edison was not about to yield control to his rival, however. Although he did have some genuine concerns over the safety of AC power, he was also motivated by a proprietary sense about the uses of electricity. His passions drove him to some rather extreme tactics in the burgeoning rivalry. He publicly electrocuted animals to demonstrate the dangers of AC power. He suggested that the state of New York adapt electrocution as its primary means of capital punishment. It was Edison's idea to use the term "Westinghoused" to describe death by electrocution.

Even so, the superiority of AC power was too obvious to ignore. Though Edison himself never publicly admitted it, the decision was made for him. By 1895, the huge Niagara Falls power facility opened, and the electricity generated there was shipped all over the northeastern United States. Edison went on to create still more remarkable inventions, but the competition between AC and DC power was one battle the famous inventor was not able to win.

Modern Mistakes

Even before computers made it faster and easier, there were many ways to take a good idea and make sure it never worked. Today's ability to make things bigger and near instant means that the failure can be more spectacular than any time in history. Many otherwise brilliant men have worked hard to give us extravagant failure worthy of these high tech times.

A Bridge Too Thin

Teresa Patterson

On July 1, 1940, a crowd of 7,000 people gathered on the banks of Puget Sound to celebrate the grand opening of the third longest suspension bridge in the world. Many marveled as the thin, graceful, 5,939-foot-long ribbon of steel undulated gently over the waters of the Puget Narrows. Called the most beautiful bridge in the world, the Narrows Bridge was a dream come true for the residents of Tacoma and the Olympic Peninsula. After trying for a bridge since the 1920s, through years of bureaucratic, financial, and design issues, and two more of construction, they finally had an efficient way to get between Tacoma and locations on the peninsula. Before the building of the bridge, the only way to reach the peninsula was to take a very long road trip around the southern end of Puget Sound, or a boat. The trip from Tacoma to Gig Harbor, previously 107 miles by car, was only eight miles via bridge. So what if the bridge rippled a little in the wind? It simply made crossing that much more interesting—sort of like a drive-on roller coaster. Few worried much about the motion of "Galloping Gertie." After all, the bridge had easily survived a 6.2 earthquake during construction, and was designed to withstand winds of up to 120mph.

Four months later, on the morning of November 7, "Galloping Gertie's" undulations suddenly turned violent during a wind storm, whipping up and down and even twisting sideways with increasing frenzy. The madly tilting roadway tipped Ruby Jacox's van onto its side. Cars slid uncontrollably across the deck as people on the bridge struggled out of their vehicles, stumbling, holding onto the curb, and in some cases crawling to escape the writhing span. Then, at a little after 11:00 a.m., with a metallic shrieking wail, the bridge suddenly ripped apart, sending the center section—and Leonard Coatsworth's car with his dog still in it—into the swift flowing waters 195 feet below. Amazingly, no humans died.

Why did it fail? On paper, the bridge was state of the art, designed by one of the most respected bridge architects of the time, Leon Moisseiff. Suspension bridges were not new. They had been around for centuries, though they had only come into their own with the advent of steel and superstrong cable in the late nineteenth century. And the Narrows Bridge could not even claim the longest span. Two other bridges, the Golden Gate Bridge in San Francisco and the George Washington Bridge in New York City, were longer. It was certainly not the widest; having only two lanes and a sidewalk, in fact it was the thinnest. The same delicate thin roadway that gave the bridge its elegant look was also its fatal flaw.

In most non-suspension bridges wind is not really a significant factor. But suspension bridges are different. Their suspended deck acts like an airfoil, reacting to the wind and creating drag and lift, especially if there are no openings to allow the wind to pass through. If the bridge deck is too flexible, it will move, lifting and falling as the wind deflects off the leading edge in much the same way an airplane wing lifts a plane. If the wind creates just the right eddies and currents, the deck begins to oscillate. It doesn't necessarily even require a lot of wind. Speed is not as great a factor as frequency. If the wind frequency matches the frequency of the moving bridge, as it did in the narrows, it creates harmonic oscillation. Once that happens the movement of the bridge starts actually building on itself, no longer dependent on the wind, creating greater and greater movement, until the bridge eventually tears itself apart.

In the early nineteenth century, wind was responsible for at least ten suspension bridge failures, though no one really understood why. They did understand that thicker and stiffer bridges seemed to do better against the wind. But Moisseiff forgot all that in his rush to create thinner, more beautiful—and cheaper—spans.

He designed the Narrows Bridge using a revolutionary solid eight-foot plate girder deck, instead of the more common thick open truss. Moisseiff based his design on the "deflection theory," that the "dead weight" elements of the bridge—the cables and deck—would create so much stabilization against the stresses of wind and traffic that there was no need for stiffening trusses or cable-stays, thus allowing much lighter, thinner spans. Of course the deflection theory was originally formulated for concrete arch bridges—not suspensions.

Despite his fame, the Narrows Bridge was the first bridge that Moisseiff designed himself. While he had worked as a designer on many of America's long suspension bridges, he had never actually been the primary designer. Moisseiff was not even the Washington State Toll Bridge Authority's architect of choice. The most spectacular bridge collapse in history might not have happened at all if they had been allowed to use their own architect.

Lead project engineer Clark Eldridge originally created a design for the bridge, utilizing six reinforcing braces on the towers and a twenty-five-foot-thick open truss system on the deck. But Eldridge's design had an estimated cost of $11 million. Moisseiff told the financiers that he could cover the same span for $6.4 million. In the end, in order to get the money to build the bridge, the State Toll Bridge Authority was forced to use "eastern consulting engineers"—which meant Moisseiff. Eldridge was still the lead project engineer, but he had to build Moisseiff's design.

In addition to the plate-girder deck, Moisseiff increased the length of the center span while decreasing the amount of support braces for the two towers. The bridge would be lighter than Eldridge's design—by hundreds of tons of steel—and narrower than any bridge built up to that time.

Eldridge did manage to retain one design element, however. Before construction began, the contractors complained that Moisseiff's design for the two support piers was "impossible to build." To keep the project moving, Eldridge ended up using his own original design for the piers in place of Moisseiff's. After the collapse, those support piers were the only part of the bridge that survived unscathed.

The new Narrows Bridge, built ten years and lots of wind tunnel and vibrations tests later, is almost identical to the design first proposed by Eldridge. But if the collapse had not occurred, engineers might never have done the research into aerodynamics, vibrations, and wave phenomena that led to a new era of stable, safe suspension bridges.

As for "Galloping Gertie"? Her sunken remains lie where they fell, at the bottom of the Puget Sound Narrows, one of the world's largest man-made reefs. In 1992, they joined the National Register of Historic Places to protect them from salvagers.

Goldie the Goldfish's Really, Really Big Cousins

Douglas Niles and Donald Niles, Sr.

The complicated relationships between people and animals go back many thousands of years. Humans have domesticated creatures ranging from birds to mammals to fish, primarily for food, but also for transportation (horses and oxen), clothing (sheep), protection, assistance, and companionship (dogs in their many varieties). When people migrated around the planet, they often brought their animals with them. The Spaniards' transportation of pigs and horses to the Americas, for example, changed the New World in ways that are still being realized today. (Horses had existed in the Western Hemisphere in prehistoric times, but they, along with most other large mammals, became extinct about the time the first humans spread across the two continents.)

However, the transporting of animals from one ecosystem to another does not always go according to plan—in fact, it might be stated that it *never* goes according to plan. The introduction of rabbits, as pets and food, to Australia backfired when the cute little bunnies (who have a rather well known penchant for reproduction) virtually overran the land Down Under. The Everglades in south Florida are being infested with pythons and other constrictors that are not native to the area. Apparently a large number of Floridians eventually decide that their pet snake has grown too big to keep, so they release it into the wild where it proceeds to devour any and all local fauna, none of which has evolved to regard these snakes as natural predators.

You'd think we would learn, but no, not us humans. Take the carp, for example. These large and prolifically breeding fish have been domesticated in China for more than a thousand years. Some breeds of carp were imported to the United States during the 1800s. Large, fast-growing, capable of eating just about anything, they took hold in many bodies of fresh water. Regarded by some as at least tolerably edible (at least, after smoking the meat) they are considered by others as merely a nuisance fish. Still, they have been around for more than a hundred years, and have become a fixture in many American lakes, ponds, and rivers.

Yet it is a different carp that is at the root of a new and very real problem currently menacing the Great Lakes—the largest fresh-water eco-system in the world. During the 1970s, two new (to America) breeds of carp were imported in small numbers by the operators of fish farms in the south, especially in Arkansas and adjacent states. Bighead and silver carp were brought in because they serve a useful economic function: they essentially worked as organic filters, cleaning the water of the fishponds by feeding on algae and absorbing other suspended matter.

They did their work well, too, and the presence of these

large carp aided the operators—most of whom raised catfish—in keeping their water clean and enhancing the efficiency of their farming operations. Things hummed along pretty well for about fifteen years. Less well known, but accurately documented, are efforts by the federal government to use these same fish to help clean up sewage containment ponds (clear evidence that the fish can survive in some really polluted water!). Small numbers of the exotic carp were transported throughout the south by federal wildlife and natural resource management officials. As in the catfish ponds, the carp not only survived, they thrived. They did their work well, and reduced the use of chemical treatments required. All in all, it seemed like a win-win situation.

During the 1990s, however, a series of catastrophic floods swept through the areas where the fishponds had been established. The farmers faced tremendous losses, but so did the environment of the whole Mississippi River system: the floodwaters carried away the borders around the fishponds, and introduced bighead and silver carp to the wild. Those fish liked the waters of the Mississippi and its tributaries—they liked them a whole lot. So much so that they have already overwhelmed many of the native fish in the Big Muddy, with devastating effect on a large number of fish-based economies.

Carp eat voraciously—up to twenty percent of their body weight in plankton, daily—and they breed with similar enthusiasm, females carrying as many as five million eggs at a time. They can grow to up to a hundred pounds, and lengths of four feet. As they continue to expand their range, they have moved northward through the Mississippi and into important tributaries including the Missouri, Ohio, and Illinois Rivers. There is every indication that the carp invasion will continue up the adjacent waters until the exotic, imported carp have taken over the rivers, exterminating or endangering countless species of native fish.

Although their consumption of plankton—which is at the base of the food chain for virtually all aquatic species—is a massive threat, the carp are not just deadly to fish and the fishing industry. An unusual, and dangerous, trait of the silver carp has rendered them a menace to the sport-boating industry, as well. It seems that the fish panic when the waters are disturbed by a loud, unnatural sound like, say, the rotating propeller of an outboard motor. The fish's reaction is to fling itself upward, out of the water, sometimes as high as ten feet in the air. When a twenty-pound fish leaps into the air directly in the path of a speeding motor boat, Jet Ski, or water skier, the resulting collision is a bruising experience, or worse.

The silver carp have become so thick along one stretch of the Illinois River that the area's whole culture has changed. People who grew up using the river as a primary focus of recreation are now pulling their boats out of the water and keeping their kids on dry ground. A U.S. Fish and Wildlife biologist who patrols the river has had a protective cage placed around the pilot's station of his boat, and is thinking about getting lacrosse helmets as protection for his crew. A fisherman who, a decade ago, might collect 5,000 pounds of native buffalo fish recently reported a haul of more than 6,000 pounds of bighead carp, and only one single buffalo.

Nature has divided the Great Lakes and the Mississippi Basin into two separate watersheds. A subcontinental divide extends from Minnesota through Wisconsin, Illinois, Indiana, Ohio, and eastward, dividing the drainage from rainfall and snowmelt. Water from north and east of the divide flows into the Great Lakes and, eventually, out the St. Lawrence River. Water falling south, or west, of the divide runs eventually into the Mississippi and the Gulf of Mexico, often through important tributaries such as the Wisconsin, Illinois, and Ohio Rivers. Such watershed divides are, and have always been, facts of nature.

But once again the hand of man has set foot upon Mother Nature's neck. With the support and funding of federal and state governments, the Chicago Sanitary and Ship Canal was created long ago as a boon to business interests. The manmade channel is still used by lots of barges, carrying goods ranging from coal to timber to food. And—you guessed it—it connects the two watersheds, joining the Chicago River, which flows into Lake Michigan, to the Illinois River, which of course flows into the Mississippi.

Right now, the bighead and silver carp have advanced up the Illinois River to within 50 miles of Lake Michigan. That huge body of fresh water, with its easy connections to the rest of the Great Lakes system, would be an environmental paradise for the carp. However, it is currently a financial and sporting paradise for millions of boaters, as well as sport and commercial fisheries. The activities of all of these users are drastically threatened by the imminent invasion.

The federal government, as well as the various affected state governments, have begun to take notice. However, the canal is still too useful, and too integral a part of the existing industrial infrastructure to consider filling it in. Instead, the feds have come up with a different plan, an electric barrier across the canal. A series of cables are used to charge the water with a jolt of electricity that, in theory and through testing, has been show to turn back migrating fish. The plan—or at least, the hope—is that this barrier will prove impermeable to fish, and will hold the invasive species out of the vast freshwater treasure of North America.

However, like so many inventions of humanity, this plan that, admittedly, looks good on paper, has already demonstrated some weaknesses. It was completed and intended to be activated in 2005, but two years later federal engineers are still trying to work out the bugs. It seems that the barrier electrifies a lot more of the water than previously imagined, and this has raised concerns about the safety of barge workers who would potentially have to unload coal in an area of sparking, charged water. Sparks have also flown from barges in transit, and the potential arc between a metal hull filled with coal and one carrying gasoline doesn't have to be spelled out to be frightening.

Furthermore, the power supply that would maintain the electrical barrier is subject to the same vagaries as the power supply of the rest of the Midwest. While it is generally reliable, the existence of tornadoes, ice storms, and, yes, floods, can all create interruptions in the supply. If the power fails, and a few exotic carp—carrying five million eggs apiece—meander up the canal and into Lake Michigan, the results, and the resulting catastrophe, are not too difficult to imagine.

Overcooking with Atoms

Teresa Patterson

The third worst nuclear disaster in history began as a brilliant plan to create clean, cheap energy without fossil fuels. In April of 1957, unknown to most Americans, the Sodium Reactor Experiment came on line at the Santa Susana Field Laboratory, 30 miles north of Los Angeles, California. The SRE was only one of ten nuclear reactors located at the SS Field Laboratory, but it was the most powerful, at 20 megawatts, and the only one designed to provide energy for civilian use. On July 12, 1957, the SRE, built and run by the Atomics International branch of Rocketdyne, began feeding energy to the power grid, becoming the first commercial nuclear power reactor in the United States. At the time, one Los Angeles paper proudly proclaimed, "L.A. Housewives Cook with Atom."

For more than two years the SRE supplied power through Southern California Edison to over 1,100 homes in the Moorpark area of Ventura County. Then in 1959 that same plant malfunctioned, causing the third worst—and least known—nuclear accident in the world, after Chernobyl in the U.S.S.R. and Sellafield in Britain but well before Three Mile Island.

After World War II, scientists were determined to harness the power of the atom for peaceful purposes. In the mid fifties, the Atomic Energy Commission began a Five-Year Reactor Development Program authorizing the design and construction of a number of nuclear reactors. The most common type, the water-cooled reactor, used uranium rods as the fuel source, controlling the reaction with water. But water reactors eventually consume the uranium fuel and must be re-supplied. At the time, fuel grade uranium was quite rare. The AEC knew that existing supplies would not be sufficient to fuel all of the reactors in their plan. They were also concerned about the difficulties involved in disposing of spent fuel. They needed a different kind of reactor.

Cooled by liquid metal instead of water, the SRE, a graphite-moderated, liquid sodium metal cooled power reactor, represented an innovative solution to the depletion problem, and an evolution in reactor design—at least on paper. Sodium reactors never deplete their fuel supply, so the SRE could—theoretically—provide uninterrupted power without any need for additional uranium. It was therefore vastly more efficient than a comparable water-cooled reactor.

Unfortunately, molten sodium is also vastly more dangerous than water. It burns in the presence of air and explodes in the presence of water. To prevent air contamination, the SRE design included a barrier of helium enveloping the reactor core. To prevent water contamination, the pumps that circulated the liquid sodium through the core were lubricated with tetralin, an organic chemical. However, since the SRE was "experimental," the one thing it didn't have was a protective shield. The Atomic Energy Commission did not require experimental reactors to have an external containment structure. Unlike Three Mile Island, with its heavy reinforced concrete domes, the Sodium Reactor Experiment resided in a common industrial building. The design included holding tanks to capture any escaping radioactive gases—but the holding tanks vented into open air. It was assumed that the delayed discharge from the holding tanks and the remote location of the facility would provide adequate protection.

But such optimistic design parameters were not the reactor's only flaw.

On July 13, 1959, during regular operations, the temperature and radiation readings on the SRE suddenly spiked. Technicians fought to shut the reactor down as readings skyrocketed—indicating the reaction was out of control. After considerable effort they managed to shut it down, but no one could explain why they had experienced the sudden "power excursion." The technicians carefully inspected the reactor—for about two hours. After that very basic inspection they still had no idea what had happened—so they did what Homer Simpson would do—and turned the reactor back on.

In true *Simpsons* fashion, Atomics International continued to operate the SRE, despite major temperature and radiation fluctuations, for two more weeks—until July 26, when radiation levels and other signs of major trouble escalated to the point that the scientists could no longer ignore the problem. They finally decided to shut the reactor down and get to the bottom of the situation.

During the next, much more thorough, examination, technicians dropped a camera into the reactor core and found the bottom—literally. Thirteen out of forty-three fuel rods had melted and fallen to the bottom of the reactor chamber. The core had been in the process of melting for fourteen days—the longest nuclear accident in history.

The scientists discovered that a leak in the pump seals caused the accident, allowing tetralin to leak through the seals into the molten silicon and then into the reactor core. Of course tetralin had been chosen as a lubricant because it didn't react with liquid sodium. But no one thought to test it inside a reactor.

Once in the core, neutron bombardment caused a change in the tetralin, turning it from a liquid into a gluey tar-like substance, eventually creating a build up of carbonaceous material that blocked coolant channels around the fuel rods. Without coolant, the fuel overheated, and eventually melted. By the time the reactor was finally shut down, one third of the fuel had been damaged.

The melting process released radioactive by-products that breached the helium layer, entered the holding tanks, then, over the course of weeks, vented into the environment. The actual extent of the release is unknown, though some monitors in use at the time went off the scale. While no reliable measurements were taken from the liquid sodium, ratios of volatile radio nucleotides found in the coolant suggest the release was significant.

The actual SRE accident was only about one one-hundredth the size of that of Three Mile Island, but experts estimate that the SRE accident released between 260 to 459 times as much radiation, making it the worst nuclear disaster in the United States. Without Three Mile's concrete containment domes, there was nothing to stop the resulting radiation from escaping directly into the air.

In October 2006, a special advisory panel of scientists and

researchers from across the U.S. concluded that the cesium-137 and iodine-131 that escaped may have caused as many as 1,800 cancer deaths among workers at the facility and civilians in the surrounding area.

Today sodium-graphite reactors are running or under construction in many parts of the world, some of them with long and successful work histories—but all of them are designed with containment shields. It is to be hoped that they also use a different pump lubricant—and have smarter technicians.

"Whoever concerns himself with big technology, either to push it forward or to stop it, is gambling in human lives."

—Freeman Dyson

Contamination in the Hills

The Santa Susana Field Laboratory

Teresa Patterson

Near the close of World War II, Werner von Braun's V2 rocket made a profound impact on the world as it streaked through the skies on its mission of destruction. Von Braun's rocket inspired future generations to reach for space, heralding the beginning of a global race for military and technological superiority that continued into the cold-war era and culminated with a drive to control the heavens. After the U.S.S.R. shocked the world with the launch of *Sputnik* in 1957, the drive to attain a foothold in space reached frenzied proportions, especially in America. The U.S. soon dominated the resulting "Space Age," demanding more and greater advances in aerospace, nuclear, and military science. Even ordinary citizens were caught up in the excitement of the quest to reach the moon. Companies such as the Rocketdyne division of North American Aviation took up the challenge, developing rocket engines, military systems, and even nuclear power reactors to quench the hunger to go higher, faster, and stronger—eventually sending American astronauts to the moon on the backs of their mighty Vanguard rocket engines, as well as putting the first nuclear reactor in space.

World War II also revealed the terrifying power of the atom. The U.S. proved it could use nuclear power for war, but the aftermath of the war launched a desperate desire to harness the atom's awesome energy for peace. Rocketdyne's subsidiary, Atomics International, joined the elite list of companies working with the government to make that dream a reality.

Of course rockets do not leap wholly formed onto their launch pads, and splitting the atom—especially when you don't want an explosion—is tricky stuff. Rocketdyne needed a place large enough to build and test rocket engines along with the new fuels to drive them, and remote enough to allow cutting edge nuclear research—since they intended to conduct work too dangerous for a populated area. In 1947, they chose a site with over 2,500 acres of wilderness high in the Santa Susana mountain range. It fit the plan perfectly. Situated between the Simi and San Fernando Valleys, it was remote enough to keep secrets, but not too far from the Rocketdyne headquarters in Canoga Park, California.

Both NASA and the Department of Defense were thrilled at the prospect of testing new munitions and propulsion systems, while the Department of Energy saw the new lab as the perfect place to explore the energy potential of the atom. At the time, it didn't seem to matter that the site was only thirty miles from Los Angeles—one of the most densely populated cities in the country. All that mattered for the government and Rocketdyne—and for much of the country for that matter—was success at any price. No one had yet considered what that price would be.

The Santa Susana Field Laboratory's state-of-the-art design included four distinct areas:

Areas I and II, partly owned by first the Air Force, then NASA, included rocket and missile labs, test firing areas, the Advanced Propulsion Test Facility, and an open air burn pit.

Area III included a systems test area and laboratories.

Area IV, partly leased by the Department of Energy and run by the Atomics International branch of Rocketdyne, contained the nuclear energy research, fuel development, and disposal facilities. In later years Area IV hosted Star Wars Laser research, also known as the Strategic Defense Initiative.

To protect any eventual human habitation—and the secrecy of the projects—a large buffer zone of undeveloped wilderness extended around much of the site.

The design seemed to take everything into account—except the very nature of the volatile substances involved and the fact that Rocketdyne was determined to "push the envelope," without seeming to consider the rule that every good cook knows—cleaning up the mess is part of the job.

In 1948, Rocketdyne began testing rockets and test firing rocket engines in areas I, II, and III. Over the years it is estimated they conducted approximately 30,000 open-air rocket tests, many successfully arching into the California sky. They also had some spectacular failures. It was not unusual for a rocket to blow up on the pad if something was out of balance. In fact, the whole point of the lab was to make certain that any failures occurred on the test pad—not during an actual launch. Rocketdyne's track record speaks for itself on that count. Their credits include the Saturn V, workhorse of the

Apollo Program, the Jupiter, Redstone, and Delta as well as engines like the RS-27A, used on the Atlas ICBM, and as the Space Shuttle Main Engine.

Of course rockets and rocket engines require rocket fuel— high grade rocket fuel, as well as oxidizers, solvents, and other volatile components necessary for getting the large chunks of metal off the ground. At the very least, setting these mega-candles off created serious emissions. At worst, in the case of spills or engine misfires, significant amounts of hazardous chemicals were released into the environment. At SSFL, the goal was getting the engines to work—no one was particularly concerned about what happened to the chemicals involved. As a result, each and every test released some degree of toxins into the environment. But that was only the beginning.

After each launch and each engine test, the test stands, pads, and engines had to be cleaned with trichloroethylene (TCE)—in copious amounts. Known to cause damage to the human nervous system, liver, lungs, and heart, TCE can cause coma and death when inhaled or ingested in large quantities.

No one knows exactly how much of the rocket fuel or cleaning chemicals leached into the ground over the years, but recently the California Department of Toxic Substances Control (DTSC) detected dangerous levels of perchlorate (from rocket fuel) in an artesian well at the Brandeis-Bardin Institute, a Jewish education camp, approximately a mile from the lab. The well showed perchlorate concentrations of 140 to 150 parts per billion (ppb). According to DTSC the highest acceptable levels by California standards are 4 ppb. Rocketdyne denied the possibility of contamination—yet they purchased Brandeis-Bardin land, now part of the perimeter buffer around the SSFL site.

The DTSC believes that even more of the chemicals are

trapped between layers of rock deep underground. Those toxins are stable—for the moment. An earthquake, common in California, could quickly change that.

Yet, despite years of fuel and chemical spills, the rocket propulsion sections of the SSFL are practically immaculate compared to the rest of the lab.

The fourth—and most secretive—section of the lab became infamous for the worst nuclear accident in U.S. history when, in 1959, the Sodium Reactor Experiment (SRE) experienced a meltdown, releasing radiation into the air (see Overcooking with Atoms). But the SRE was not the only nuclear reactor on site, though it was the largest. Over the years a total of ten nuclear reactors, mostly low power, operated out of area IV. One of them, the SNAP8DR, became the first nuclear power system in space, launching from Vandenberg AFB on April 3, 1965. To this day SNAP8 remains the only nuclear reactor placed in space by the United States.

The Sodium Reactor was also not the only reactor to have a nuclear accident. At least four of the ten reactors experienced accidents. In addition to the 1959 SRE meltdown: the AE6 reactor experienced a release of fission gasses the same year; in '64 the SNAP8ER experienced damage to 80 percent of its fuel; in '69 the SNAP8DR (new and improved) received similar damage to one third of its fuel. Lesser radiological incidents within the reactors included a wash cell explosion and a steam cleaning pad contamination incident.

That many accidents—especially the radioactive kind—would have been bad enough. But it got worse. Because all the reactors were classed as "experimental," none of them had containment structures. Any vented radioactive materials were free to escape directly into the environment.

And yet these reactor accidents were probably not the worst events to happen at the SSF lab.

In addition to the reactors themselves, area IV had an entire array of nuclear support facilities. Known as "critical facilities," these included a facility for fabricating plutonium fuel rods; a uranium-carbide fuel fabrication facility; a sodium burn pit, in which non-radioactive sodium-coated objects were burned clean in an open-air pit; and a "Hot Lab."

The Hot Lab, a facility for remotely cutting up irradiated nuclear fuel, specialized in the disassembly and on-site storage of nuclear material. The Atomic Energy Commission and the Department of Energy both shipped spent nuclear fuel from their facilities around the country to the SSFL lab to be decladded and examined. Thought to be the largest Hot Lab in the country at the time, it was probably not the safest—though it was definitely "hot." The lab suffered several fires involving radioactive materials. In 1957 a fire inside the Hot Lab got out of control and according to reports at the time "massive contamination" resulted. Passage of time did little to improve Hot Lab safety. As late as 1971 documents record a dangerous fire involving contaminated combustible reactor coolant.

Many environmental researchers believe that accidents at the Hot Lab and the Plutonium Fuel facility may have been more serious than even the SRE meltdown, but very few details about their accident histories have been made available to the public.

Yet while activities at both the propulsion labs and the nuclear facilities created potentially lethal environmental hazards to both workers and nearby civilians, it was the scientists on "trash detail" who faced the greatest immediate danger.

Between the propulsion lab chemicals and the nuclear components, the SSFL generated large amounts of hazardous material waste, much of it quite volatile or toxic. In addition, toxic and radioactive materials were regularly shipped in from other facilities. Yet, for some reason, despite all the other state-of-the-art facilities, the SSFL design omitted any equally high-tech way to dispose of all the hazardous waste. Most of it ended up in barrels, and the storage space was limited. Rather than attempt to continue storing the rapidly accumulating barrels of poison until the facility was buried in it, the order came down to dispose of it. So the lab workers did.

Utilizing a unique—and possibly insane—method of disposal, the crew responsible for disposal duties placed barrels marked for disposal in an open area, then, from behind the protection of a small barrier some distance away, they took a rifle, took aim—and shot each barrel. The resulting explosion, caused by the bullet impacting the volatile contents, usually did succeed in destroying the offending barrel. As for the contents? It appeared to have been vaporized, but in fact much of it simply became airborne—a true case of out of sight, out of mind.

It is unclear exactly when this practice began, but according to DOE official Mike Lopez, as reported by the *Ventura County Reporter* in 2002, it was used as a typical cleanup procedure for a number of years, only ending sometime prior to the 1990s. The California legislature even found it necessary to create a bill (SB990) concerning this manner of waste removal.

Target shooting, while unique, was not the most dangerous method of waste removal. Because of the large number of nuclear operations that utilized liquid sodium, the SSFL had an open-air pit, known as the sodium burn pit, for cleaning sodium-contaminated components. On paper, this pit was only for burning off the sodium buildup to allow the components to be reused. In practice, however, it was used as a trash-burning pit for everything—including chemical and radioactive waste. One worker, James Palmer, interviewed by the *Ventura County Star,* spoke of going home at night to kiss his wife, only to burn her lips from the chemicals that accumulated on his own lips while at work. The article went on to report that out of the twenty-seven men on his crew, twenty-two died from cancer.

On July 26, 1994, the dangerous practice turned deadly. Otto Heiny and Larry Pugh, both scientists at the facility, were under orders to burn the contents of unmarked containers in the burn pit. Believing they were burning the leftover chemicals as allowed by law, Heiny poured an unmarked container of explosive material into the fire. Both men were killed in the resulting explosion, and a third man injured in the blast. When the case went to court in 2004, three Rocketdyne officials pled guilty to illegally storing explosive materials. The more serious charges, related to illegally burning hazardous waste, resulted in a deadlocked jury. Rocketdyne only recently admitted that toxins, including napalm, dioxins, and unmarked waste products from the nuclear facilities were regularly burned at the burn pit.

Even when used successfully, the burn pits released toxins into the environment in the smoke from the burning and in toxic runoff every time it rained. Rocketdyne claims to have sealed the pits with impermeable clay to prevent runoff—but it turns out the clay was actually very porous, only slowing the runoff.

In 1989, a DOE investigation found widespread chemical and radioactive contamination throughout the property. The resulting public outcry at this revelation led Rocketdyne to terminate all nuclear activity at the site. A number of lawsuits and the beginning of cleanup operations followed.

In 1995, the EPA and the DOE entered into a joint agreement to clean the site to EPA Superfund standards. Boeing purchased

Rocketdyne 1996, inheriting the entire mess. However in 2003 the DOE and Boeing, probably realizing the true scope of the cleanup, performed an about-face and announced the SSFL property would no longer be brought up to Superfund standards, but instead put forward a plan that only involved cleanup of 1 percent of the contaminated soil and included a plan to release the land for unrestricted use in approximately ten years. The resulting battle between the EPA, the local residents, and Boeing/DOE is still raging.

It is doubtful the advances in aerospace technology that led to the current space program would have been possible without the vital contributions from the Santa Susana Lab. Nuclear science, too, owes a great debt to the work done at the lab. But because Rocketdyne focused so completely on attaining these advances without giving serious thought to the consequences, the employees of the lab as well as the people of southern California may pay the price for decades to come.

In August of 2005, the Pratt & Whitney Corporation purchased Rocketdyne from Boeing. There was one condition, however—the Santa Susana Field Laboratory could not be included in the sale. As of this writing Boeing still owns the SSFL.

> "Believe it or not, I'm walking on air. / I never thought I could feel so free."
>
> —Mike Post and Stephen Geyer, theme song from *Greatest American Hero*

Under Pressure

The First Space Walk

Teresa Patterson

Soviet cosmonaut Alexei Leonov was the first man to view the awesome vistas of space unencumbered by the confines of a capsule. Now an artist, he has spent much of his life trying to capture the wonders of those moments on canvas. But Leonov came very close to becoming a permanent part of that vista.

In 1965, both the U.S. and Soviet manned space programs were fighting to become the first to land a man on the moon. At that point, the Soviet Union led the race, having launched Sputnik 1, the first successful man-made satellite to orbit the earth, followed by the first manned orbit, with cosmonaut Yuri Gagarin, and the first woman in space, Valentina Terechkova. Russia had even scored the first three-man space mission, though they had to send the men up without spacesuits in order to fit them all in the tiny Voskhod capsule. But the Russians felt the

United States closing fast on their heels after the success of the Mercury program and preparations to launch the first Gemini mission.

To hold onto their lead, the Soviets planned to prove a man could function successfully outside his space capsule and "walk in space." It seemed a simple enough objective. They had already proven they could get men into space and back down again. All they had to do was get a man out of the pressurized capsule—without having his blood boil in the vacuum of space—and get him back in again a few moments later.

Of course, as one poet observed in a song popularized at sci-fi conventions, if you open a pressurized spaceship hatch to the vacuum of space, "the air is sure to leave it. And air is very hard to catch, you never will retrieve it."

In order to prevent the depressurization of the entire capsule, the Soviet designers created an "airlock," a small isolated area with an inner and outer hatch that would allow controlled depressurization without endangering the pressure of the entire cabin. To keep the airlock from taking up valuable space in the tiny cabin, they made it inflatable.

To protect the cosmonaut from the deadly vacuum, they created a strong, flexible space suit, a human-shaped pressure balloon, which would hold a high-pressure artificial atmosphere around his body.

On March 18, 1965, Alexei Leonov and Pavel Belyayev rode their Voskhod 2 spacecraft into orbit. Approximately an hour later while Belyayev piloted the craft, Leonov inflated the airlock, entered it, and stepped through the hatch into space. The young cosmonaut spent twelve minutes making history as he floated outside his ship, anchored only by a five-foot tether.

During those twelve minutes, however, Leonov's flexible

suit responded to the lack of external pressure by expanding—
much the way a balloon does as it rises in the atmosphere. It
also became rigid. Upon his return to the ship he discovered he
could no longer fit back through the tiny hatch. He found himself trapped outside his ship. In order to get through the hatch
he opened a valve and began bleeding air from his suit, a little at
a time. After almost ten minutes he still wasn't small enough to
get all the way into the airlock. Finally, in desperation, he bled
off almost all of the air from his suit and still barely managed to
struggle, feet first, into the airlock, and, with effort, sealed the
outer door. He made it just in time for the ship's passage over the
ground stations in the Soviet Far East.

Unfortunately Leonov's problems didn't end there. Notori-
ous for technical glitches, the automatic guidance control malfunctioned, omitting a crucial command to initiate the reentry
sequence. Pavel Belyayev had to land the ship manually with instructions from ground control. For some reason, during sixteen
revolutions of the orbit, ground control kept sending the order
to initiate retrofire after the ship had already passed the retrofire
point. Belyayev managed to bring the capsule down safely—but
landed over one thousand kilometers off course in a snow bank
deep in the Ural Mountain wilderness. The two cosmonauts
were forced to spend the night where they landed while rescue
crews struggled through the thick forest to reach them. Even
then, the area was too heavily wooded to allow a helicopter to
land, so they had to ski to a safer landing zone before being airlifted back to civilization. The two cosmonauts finally returned
to the Baikonur Cosmodrome approximately forty-eight hours
after their initial landing, having spent more time in the wilderness than they had in their entire historic mission.

The Biospherians in the Bubble

Joshua Spivak

Science has never been able to fully answer the question of why the earth is well suited for life. James Lovelock, author of the Gaia hypothesis, argued that the living organisms on earth evolved in symbiosis with their surroundings. But how could this theory be tested?

Lovelock and other scientitsts hatched an idea.

What about a working model of our world?

They built a miniature, completely self-contained version of earth—in Arizona—filled it with a diverse array of species, and locked in eight scientists as well, serving as the "keystone predator" in this quasi-laboratory, for a stint of two years.

If they were correct, the air, water, and life forms would work symbiotically.

If it didn't, they would at least have a wealth of scientific data.

Rather than blow everything on an "all or nothing" venture, the plan prescribed a series of small qualitative and quantitative steps. The first was extrapolated from a series of experiments performed by Russian scientist Evgenii Shepelev, in which he constructed and inhabited a completely sealed steel chamber filled with only green algae in a water solution for a trial period of twenty-four hours (showing that the algae's production of oxygen combined with his production of carbon dioxide balanced each other out). A group of American scientists next locked themselves in an enclosed environment (one that grew food and had water condensed from the humid air) for three weeks.

With each test verifying the scientists' theories, by the late 1980s they were ready to re-create the project on a large scale. Funded by Edward Bass, the scientists built an eight-story complex, costing one hundred and fifty million dollars, in the Arizona desert, and named it Biosphere 2 (Earth being the original Biosphere). Once the complex was constructed, Biosphere 2 was stocked with living organisms, including four thousand different species, and a number of different environments.

On September 26, 1991, following a ton of generally positive press, an eight-member team of "Biospherians," all dressed in dark blue *Star Trek*-style uniforms, were sealed into the Biosphere for two years. (Unbeknown to the public, a three-month supply of food had been stored in the Biosphere before it was closed.)

Shortly after the doors were sealed, the scientific underpinning of the Biosphere began to falter and fail pretty quickly. First, oxygen levels rapidly declined. Then, consequently, the Biosphere's air quality declined. By December, the Biosphere's leaders were forced to acknowledge that outside air was being pumped into the project. Eventually, the oxygen levels fell from twenty-one percent to fourteen percent—barely enough to sustain life—with a corresponding rise in carbon dioxide and nitrous oxide high enough to cause brain damage.

Outside, negative stories started popping up in the international press.

Inside, the crew faced constant hunger—and traded accusations of food theft—as the production of food did not come close to satisfying the occupants (one Biospherian went from 260 pounds to 150). The only hope for a successful conclusion to the project relied on a last ditch repair effort. But instead of fixing the project, the repairs only made matters worse.

Morning glory vines, introduced to soak up excessive carbon dioxide, grew so fast that they destroyed food crops. Several species started dying off. According to the *New York Times*, "Nineteen of 25 vertebrate species went extinct, as did all pollinators, dooming most plants to seedlessness. Most insects died off, except for katydids, cockroaches and crazy ants." The native Arizona ants, which were not included in the Biosphere, infiltrated the structure and ruled the day.

The planners had also radically misjudged the environments. An overabundance of rain ruined the desert. The temperature unexpectedly rose, and the light, supposed to be filtered in through the glass panes, was too dim. At the end of the first year, after a hearty series of denials, the scientists acknowledged making "10,000 mistakes" and promised to improve.

But it was way too late.

By the time the original eight Biospherians left the project, everything was in disarray. Infighting among the team members grew and the board splintered. Unsurprisingly, the guy with the most money, Ed Bass, was left standing. After ousting the board, Bass chose to donate the project to Columbia University in 1996.

Columbia recognized the problems with Biosphere 2 and tried to make the best of it. It turned the entire project on its head. Rather than having a paradise that would show humanity how to live on other worlds, Columbia instead decided to turn Biosphere 2 into a scientific test of the dangers of global warming. The high temperatures and high levels of carbon dioxide would turn the dome into what a *New York Times* reporter called a "kind of atmospheric hell." But the Biosphere could not even get this nightmare dystopia right. Again the desired results failed due to technical problems.

By 2003, the university had washed its hands of the project, turning it over to new owners.

Biosphere 2 never achieved its goal of showing humanity how an enclosed world could work together. It never proved how humans could survive on another planet, but it will become what nature truly intended for the Arizona desert: a tourist trap and a planned community.

Mars or Bust

Teresa Patterson

The angry red planet named for the god of war has inspired imaginations since humans first looked upon the heavens. The dream of exploring Mars, once believed only a fantasy, became a reality within the last half-century. But so lofty a goal rarely comes easily, and to say that the Mars program occasionally deviated from the plan would be a serious understatement. In fact, it is quite miraculous that we managed to get our probes there at all.

The Mars Program plan seemed simple enough—at least on paper. First, a series of satellites would "flyby" the red planet, powered largely by solar panels, snapping photographs and taking readings as they zoomed past. Second, another series of "orbiters" would launch with instructions to stop and enter into Martian orbit, gathering still more data and relaying it back to Earth. The next phase involved landing on the Martian surface with a "descender/lander" to take readings from ground level, followed by a phase utilizing mobile robot vehicles or "rovers" that could gather additional, more detailed data. The landers and rovers would then relay that data back to the orbiters and Earth.

We had already proven that we could get a satellite into space and keep it there. We had also proven we could safely land a craft from space. All we had to do was make it happen several million miles farther away on another planet.

Unfortunately, if the Russians were any example, it wasn't as easy as it seemed on the drafting board. The Russians led the space race, with the first satellite, dog, man, and woman in space, but by 1964, despite five launches, they had still not managed one single successful Mars mission.

Marsnik 1 (also called Korabl 4, Mars 1960A), launched on October 10, 1960, would have been the Soviet Union's first planetary probe. It was equipped with two folding solar panels to provide power for its long journey as well as a vast array of sensing, recording, and communications equipment. The launch vehicle, a new rocket designated SL-6, was basically a Molniya booster with an added fourth stage.

The new rocket never even made it to Earth orbit. After reaching 120 kilometers, the third stage failed to ignite and the ship returned to Earth the hard way. The Soviets launched Marsnik 2 (Korabl 5, Mars 160B) a few days later only to watch it suffer the same fate. Apparently it didn't occur to them to try to fix the problem before turning another expensive rocket into a smoking crater.

It took two years for the Soviets to try again. Sputnik 22 (Korabl 11, Mars 1962A) was launched October 24, 1962, for a Mars flyby. The SL-6 rocket made it to Earth orbit this time, but the spacecraft exploded shortly after, probably during the engine burn that was supposed to put the ship into Mars trajectory. Pieces of the shattered ship remained in orbit for days.

Bad as it was, the failed mission almost cost the Soviets more than a ruined spaceship. The launch occurred at the height of the Cuban missile crisis. In Alaska, the United States Ballistic Missile Early Warning System detected the debris and initially mistook it for the start of a Soviet nuclear ICBM attack. Fortunately someone figured it out before pressing the red button.

The Soviets' next attempt, Mars 1 (Sputnik 23), which launched less than a week later, was almost a success. An automatic interplanetary station designed to fly by the red planet at a distance of approximately 11,000 km, Mars 1 was equipped to collect and relay data on surface imaging, cosmic radiation, the planet's magnetic field, atmospheric structure, and lots of other cool critical data. Equipped with two solar panels, this craft actually made it out of orbit and well on the way to Mars before the communication system failed, making it impossible for ground control to send course corrections—or receive the data the little probe was probably still dutifully collecting. Mars 1 eventually passed uselessly by its namesake at a distance of almost 200,000 km, ending up in orbit around the Sun. It did succeed in taking and transmitting some valuable readings from open space before the radio failed, but none of them had anything to do with Mars.

Korabl 13, launched a short time later that same year, proved that the U.S.S.R still didn't have the bugs in their SL-6 rockets worked out. In what must have been a frustrating repeat of Sputnik 22, it too exploded upon reaching Earth orbit. Fortunately, by this time the folks at U.S. Missile Defense knew how to tell the all-too-common Russian space debris from an incoming ICBM. The Russians pretended Korabl 13 had never happened at all, omitting it from most records and waiting until 1964 to try again.

The U.S.S.R. was 0 for 4, and yet NASA decided that the Martian curse only applied to Russians. The Mars program administrators were certain the superiority of American know-how would prevail in reaching the red planet. On November 5, 1964, NASA launched the first U.S. Mars probe, Mariner 3. Avoiding the launch vehicle problems that plagued the Russians, NASA used the tried and true Atlas Agena rocket, a proven military intercontinental ballistic missile, rather than attempt to design a new booster with new problems. Mariner 3 was actually the third probe of its type, though the first allocated for Mars. Mariner 1, a flyby mission to Venus, had failed, its rocket veering wildly off course shortly after launch. Mariner 2, however, had completed its flyby, successfully recording conditions on Venus. Mariner 3 was simply the same mission with a slightly different destination.

The launch was successful, the powerful Atlas lifted its payload into space as planned—but the probe went nowhere. Its protective shroud stuck, failing to jettison, trapping the probe and ending the mission. It was NASA's first encounter with the "Mars curse," but not the last.

The next launch, later that same month, had better luck. Mariner 4 became the first probe to complete a Martian flyby, take readings and photographs, and actually transmit them back to the Earth.

Meanwhile, the Soviets launched their fifth (or fourth, depending on who's counting) attempted Mars mission, Zond 2. But this one, an ambitious design, was an automatic interplanetary station fitted with six experimental low-thrust plasma ion engines that allowed it to maneuver without expending fuel. Zond also carried a special descent craft for entering the Martian atmosphere. Otherwise Zond contained all the same basic equipment that had been on board the ill-fated Mars 1.

On launch day, the Tyazheliy Sputnik launch delivery vehicle made it to parking orbit—without blowing up—and successfully released Zond 2 to unfurl its solar panels—well, one of them anyway—and head on its merry way to Mars. Unfortunately that Mars 1 equipment had apparently seen no improvements in the intervening years. Six months and three-quarters of the way to the rendezvous, the radio stopped responding. Three months later, in August of 1965, Zond 2 passed silently within 1500 km of Mars—the closest failure yet.

In 1969, the United States launched two more Mariner probes, 6 on February 24, and 7 on March 27. The two were identical, fully automatic—although each could be reprogrammed from Earth at need—and both were successful, relaying valuable data and photos back to Earth.

At the same time, in a mission that was never officially announced, the Russians decided to one-up the Americans by moving directly to stage two, Mars orbit—even though they had yet to succeed with stage one. On March 27, the same exact day as the Mariner 7 launch, the Soviets, who routinely attempted to launch planetary probes in pairs, launched not one, but two Mars orbiter missions simultaneously.

At 10:40 UTC (Coordinated Universal Time), Mars 69A, powered by the more reliable Proton SL launcher, lifted gracefully off its pad and completed a successful burn and separation from stages one and two—before the stage three engine exploded a little over seven minutes into the flight, scattering pieces of the ship over the Altai mountains.

Nearby, Mars 69B didn't fare as well. Its Proton booster failed less than a second into the flight when one of its six first-stage rocket engines exploded. The crippled ship's control systems struggled to compensate for another twenty-five seconds before the rocket began to tip over. Then all five remaining engines shut down. Forty-one seconds into the flight, Mars 69B slammed into the ground and exploded. Designed for Mars, it managed to travel three whole kilometers.

Oblivious at the time to the failure, NASA successfully launched Mariner 6 and 7 on flyby missions, retrieving valuable images and data from both, but not before Mariner 7 gave ground control a serious scare. Months after launch, for no known reason, the probe lost communications, battery, and power and started gaining speed and tumbling erratically. Ground control watched helplessly, fearing that the curse had found their little probe. Then, equally inexplicably, five hours later Mariner 7 regained communications and power, and corrected its velocity.

Scientists started commenting, only half joking, that the "Great Galactic Ghoul" was waiting to waylay any craft that dared attempt to reach Mars. The Ghoul became the "here there be dragons" warning of the space age.

Mariner 7 suffered no more strange failures, so NASA went ahead with the next step, the Mariner Mars 71 Project. The plan involved two separate Mariner spacecraft which would insert into Martian orbit to perform different, but complementary tasks.

The plan went bad when Mariner 8's Atlas-Centaur booster experienced trouble soon after liftoff. The upper stage began to oscillate, then tumble, causing the main stage engine to shut down only a little over six minutes into the flight. The payload and engine separated before re-entering the atmosphere to fall into the Atlantic Ocean.

But NASA was undaunted. Twenty-two days later Mariner 9, now a solo mission, launched successfully and carried out its planned orbital reconnaissance, sending back stunning pictures of Martian volcanoes and canyons.

But the Galactic Ghoul wasn't through.

That same year the Russians, now determined to at least beat the U.S. on to Martian soil, decided to skip both the flyby and orbiter stages—since they had done so well with them—and move directly to the landing on the planet stage. Again working in pairs, they built two identical ships, designated Mars 2 and 3 respectively, each carrying both an orbiter and a descent/lander module. The sister ships were designed to approach Mars and enter into orbit, recording data as they went, then, once safely in orbit, to detach the lander to float through the thin atmosphere and gently touch down on the Martian surface. The landers, each proudly carrying the Soviet coat of arms, would then study the topography, soil, magnetic fields, and so forth, and relay the findings back to the orbiter.

Both ships launched successfully on May 19 and 28—using Tyazheliy Sputnik rockets this time—and approached Mars as planned. Four and a half hours out from the planet, Mars 2 released its descent module. The module entered the atmosphere—and fell uncontrollably as the descent systems malfunctioned, failing to fire braking or maneuvering thrusters. Moments later the little lander smashed—not so gently—into the Martian surface, delivering the Soviet coat of arms to the planet at the bottom of a smoking crater. The Mars 2 lander gained the dubious record of being the first manmade object to reach the Martian surface.

Mars 3 fared better. The Mars 3 lander actually managed to land relatively softly. It opened its petal-shaped protective covers and began transmitting an image to the orbiter—for all of twenty seconds. After twenty seconds it simply went dead, enveloped and destroyed by a massive dust storm. The Russians still insisted they had earned the title for the first soft landing—though the resulting shards of useless space junk left sitting in the red dust certainly didn't look like a successful surface mission.

They did finally manage the orbiting part, however. The Mars 3 spacecraft entered a creditable orbit, despite a fuel loss that forced a change in the original trajectory, and transmitted valuable data for almost eight months.

In 1974, the Russians also finally got their flyby mission with Mars 4—though on paper it was supposed to be an orbiter. A degraded computer chip failed to fire the retro-rockets, so Mars 4 never slowed down, flying right on by its target—though it did dutifully record data. The Russians got lots of information about the open space between planets as Mars 4 raced towards the sun.

Mars 5 and 6 succeeded in reaching Mars, though 5, which was intended as the main orbiting communication link to the landers carried by 6 and 7, failed after only nine days. Mars 6 successfully deployed its descent module. The lander transmitted atmospheric data almost all the way to the Martian surface before it suddenly went silent—probably after hitting the ground at around 61 meters per second. Mars 7, the last in the series, released its descent module early, so that both the ship and the lander sailed right on by the planet to be lost in solar orbit.

Feeling buoyed by the success of the Mariner mission—and

the fact that the Galactic Ghoul seemed to prefer the taste of Soviet hardware—NASA scientists took their time, and lots of money, preparing the Viking Orbiter/Lander Program. In 1976, Viking 1 made the first really successful landing on Mars— Russian space debris and coat of arms notwithstanding—and Viking 2 followed with equally glowing results. Both orbiters and landers worked almost exactly as planned, sending back thousands of images and massive amounts of data. Americans forgot about the curse and began to get comfortable with the idea of a United States presence on Mars.

The Russians couldn't forget. The ghoul got two more ships intended for a Martian moon in 1988, though Phobos 2 did send a few nice photos before it failed.

Seventeen years after Viking, the United States prepared to launch again. The Mars Observer was to be first of a series of so-phisticated observation satellites designed to study the climate, geology, topography, and gravitational and magnetic fields of Mars. Based on a commercial communications satellite design, the Observer had not two but six solar panels, a long-lasting high-tech battery system, high-resolution cameras, and state of the art systems that would allow it to map the planet from orbit with a resolution of one meter, one hundred times better than all previous photographs. It was also the most expensive plan-etary mission yet, costing over $183 million.

The Observer worked like a dream, exactly according to plan until August 21, 1993. Three days before the scheduled Martian orbit, ground control prepared the ship for firing its engines by pressurizing the fuel tanks. To prevent loss of the transmitters from the stress of the fuel pressurization sequence, ground con-trol ordered Observer to switch off all communications. After the time allocated for the prep-sequence they switched communications back on—but Observer wasn't there. They continued to try for hours, then days, but the probe, all 183 million dollars of it, had simply vanished. Even now no one knows for certain what happened to it. Experts believe it probably exploded during the fuel pressurization sequence. After all, the pressurized fuel was designed to explode when mixed. If it happened to leak and mix in the wrong place, like the titanium tubing, it would still explode—with much less pleasant results.

But Observer could also have gone on to enter orbit, its little mechanical mind blissfully unaware that it had missed the sequence to turn its transmitter back on; or it might be aimlessly wandering the solar system. In any case the Galactic Ghoul was back, with a vengeance.

Observer's disappearance cost NASA much more than a satellite. The entire program came under stringent review. Expensive missions were fine if they were successful, but not when they wasted taxpayer money for nothing. Observer was the last huge expensive probe NASA would ever launch. The Clinton administration pressured the space agency to find a new way of doing business, or stop doing business. To get approval, all missions would have to be faster, better, and cheaper, so that if one craft were lost it would not be such a significant setback.

At first, the new "better, faster, cheaper" NASA functioned like the ideal space agency, redesigning its entire approach to Mars missions. The new "Discovery" program of low-cost high-science return robotic missions, involved launching a dual mission every 18 months or so, a much tighter schedule than the space agency had ever tried to run. The next two missions incorporated much of the same science instrumentation originally flown on the Observer, but at a fraction of the cost, each coming in at under $250 million.

Both were roaring successes. The 1996 Mars Global Surveyor delivered the first high-resolution surface map complete with topographical details despite a problem with a cracked solar panel, producing more images than any other mission. The Mars Pathfinder, a lander mission, launched later the same year and actually arrived before Surveyor. Instead of braking thrusters, its lander used a cheap low-tech landing system of parachutes and airbags. Inside, it carried the first Martian rover. Sojourner immediately became a media darling as it drove around the landing zone transmitting a mobile, ground level view of the red planet. The lander was only designed to last for a month, and rover, a week, but the lander lasted more than three months. At the time the lander finally died the rover was still functional. Programmed to circle the area in the event of a communications silence, little Sojourner could be circling there still.

Also launching in 1996, the Russians tried again, this time with European collaborators. The craft failed soon after launch, crashing back to Earth, cause unknown.

NASA, riding high on their return to glory, quickly followed in 1998 with the Mars Climate Orbiter. Designed as a weather mapping satellite, it also doubled as the main support relay for the upcoming Mars Polar Lander.

Part of the newer, faster, cheaper plan required outsourcing projects previously done in house. It also reduced the budget for system redundancies and testing. No one realized that one team of programmers was working in metric while another was working in English units. The difference between the two is small in most things, but once the Climate Orbiter reached Mars, the math error amounted to 90 kilometers, the difference between traveling above the atmosphere—or in it. Orbiters are designed to orbit, not fly. The Climate Orbiter burned up in the atmosphere.

Perhaps the little Sojourner woke up the Galactic Ghoul. If so, he woke up hungry. That same year Japan launched their first Mars probe, the Nozomi, an international effort containing equipment designed by scientists from around the world. Heralded as a unique multi-nation event, it also marked the first time the Canadians participated in a planetary mission. It should have been simple. All they wanted to do was park in Mars orbit, but instead Nozomi became the ultimate Mars hard-luck story.

Trouble started moments after launch when Nozomi veered off course. The controllers managed to correct the flight path, but had to use extra fuel to do it. When the orbiter reached Mars it no longer had enough fuel to break for orbit, zooming past its target in 1999. Engineers managed to recalibrate its flight, giving it a course that would ease it into orbit more gently— at a cost of four more years and two more rotations around Earth. Then in 2002 a solar flare knocked out all power, causing the little remaining fuel to freeze. In a last ditch effort to save the ship, engineers, lacking galactic jumper cables, toggled the probe's power on and off hundreds of times as Nozomi passed near the Sun, hoping to tap into the Sun's energy to jump-start the ship. It actually worked. But the ship corrected its path only to suffer a fatal electrical short in its navigation system. In the end Nozomi barely missed Mars as it passed by into interplanetary darkness.

While the Japanese fought to save their probe, NASA, still smarting from the Climate Orbiter loss, launched the Mars Polar Lander. The lander, whose mission included searching for signs of water or ice at the Martian pole, carried two special probes, Deep Space 1 and 2, designed to crash into the planet and burrow into the ground to search for water. As it happened, it was the Polar Lander that crashed instead.

Landers usually slow their descent using special thrusters designed for that purpose. Previous landers used their radar to sense when they were ready to touch down, and thus shut off the engines. To save money and simplify things, the Polar Lander was equipped with special sensors on each of its "feet" designed to sense the upward bounce the moment it touched down, and turn the engines off. Unfortunately, later testing discovered that the very act of deploying the landing legs creates enough shock to cause the same amount of "bounce." The Polar Lander probably fired its engines, deployed its landing legs, then sensing the resulting flex, concluded it was already on the ground and immediately turned the engines off again. It would have fallen the remaining 40 meters like a stone. Budget cuts and mismanagement were cited as the reason the glitch had not been discovered in testing before the mission.

Since the 1960s there have been thirty-six Mars missions. Twenty of those have ended in disaster. Since the sixties, the Russians have only managed four qualified successes out of eighteen attempts. Yet they are not incompetent. The Soviets ruled Venus, achieving fifteen successful missions to that planet, yet Mars defeated them. Mars is also largely responsible for forcing NASA to redesign its methodology—twice. Yet we keep coming back, unable to resist the lure of the red planet.

And somewhere out there the Great Galactic Ghoul is waiting.

"If a blind man leads a blind man, both will fall into a pit."

—Matthew 15:14

Myopia in Space

Joshua Spivak

Ever since Galileo's first telescope scanned the skies, scientists have created bigger and better instruments to unlock the mysteries of the universe. But even the largest earthbound telescopes have serious limitations. While the skies seem clear, earth's atmosphere makes celestial gazing blurry. With the growth of rocket technology, in 1990, the National Aeronautics and Space Administration (NASA) launched a new type of satellite that was supposed to make our view of the heavens clearer, the Hubble Space Telescope.

Unfortunately, quality control was not in the plan.

Originally proposed in 1946, the forty-three-and-a-half-foot-long Space Telescope was designed to bypass atmospheric conditions and allow the use of equipment unmatched on earth. Placed in orbit around the earth, the astronomical satellite would take clear pictures of the outer reaches of space and time, granting scientists a window into the origins of the universe. While the goals were lofty, the projected budget did not come cheap and, with the Cold War raging, few politicians thought they could spare the funds.

In the following years, teams of scientists continued to fight for the project. Finally, in 1977, Congress approved a then-enormous price tag of $475 million for its creation. By the time the Hubble finally lifted into orbit on the Space Shuttle *Discovery* in 1990, a mere seven years late, the price had ballooned to more than $2 billion.

Still, the potential benefits were enormous.

According to astronomer Robert Smith, with the Hubble, "stars should be observed at distances about seven times greater than with ground-based telescopes." Without the distorting glare of the atmosphere, astronomers would then be able to give more accurate readings of a star's location. Additionally, images in infrared and ultraviolet light, which are generally absorbed by the atmosphere, would also be observable.

Around launch time, the scientific community was aflutter with the possibilities—but then they turned on the orbital telescope. Astute observers quickly noted a serious recursive and potentially fatal issue in the downloaded data. The pictures were nowhere near as sharp as they should have been. After running nearly every conceivable test in mission control, the scientists and technicians finally uncovered the cause of the error. One of the subcontractors, the PerkinElmer Corporation, who had underbid their contract to build the mirror so it would properly focus the incoming light, blew it.

PerkinElmer, bidding sixty-nine and a half million dollars (and eventually charged over three hundred million dollars for the shoddy work), tested the mirror by flashing a light beam through a tiny lens. For an effective test, the lens had to be set up at precisely the right distance. To align the lens, the technicians beamed a light through a rod with a special painted cap centered by a hole. Unfortunately, no one noticed that a spot of paint had worn off the cap, throwing off the grounding of the primary mirror by 1.3 millimeters. In a telescope mirror that must be able to reflect the tiniest amount of light clearly and accurately, this seemingly tiny error was potentially catastrophic.

NASA was showered with criticism, as even one of the projects' staunchest defenders, Maryland Senator Barbara Mikulski (whose state benefited from the government spending on the Goddard Science Center), called the Hubble a "techno-turkey."

Harried scientists debated a host of solutions (including bringing the telescope back to earth), eventually deciding on a seven-hundred-million-dollar space shuttle repair mission.

In December 1993, after two years of intense training, a team of astronauts flying on the *Endeavour* hand-wrestled the spinning telescope into position. Over the course of eleven days and five elongated space walks, the repair team, led by Payload Commander Story Musgrave, removed a key piece of equipment, the high speed photometer, and replaced it with a device that refocused the light coming in from the primary mirror. Despite the danger, the mission, arguably the highlight of the entire space shuttle program, went perfectly.

If not for the work of scientists and a team of astronauts, the Hubble might have gone down in the books as one of the great debacles of American science.

> "I don't know jokes; I just watch the government and report the facts."
>
> —Will Rogers

The Starr Report

Brian M. Thomsen

In September of 1998, a new publication premiered at number one on the *USA Today* bestseller list. It had all of the requisite elements of a commercial hit novel—intrigue, deceit, conspiracy, and plenty of salacious sex.

But this wasn't a novel.

It was nonfiction, and it was authored by an agent of the United States government and thoroughly authorized by Congress.

Called the Starr Report, it was designed to make the case for the successful impeachment and removal from office of the President of the United States. Its actual title was far more mundane: *Referral to the United States House of Representatives pursuant to Title 28, United States Code, § 595(c), Submitted by the Office of the Independent Counsel, September 9, 1998*, and it was the product of a several-years-long investigation into certain events involving President William Jefferson Clinton, ranging from a failed land deal in Arkansas, through an alleged sexual

harassment lawsuit, right to allegations of perjury and obstruction of justice revolving around adulterous sexual activity in
the White House.

The actual cases involved were incredibly convoluted, as was
the investigation, managed by Independent Prosecutor Kenneth
Starr, a former Solicitor General for the Reagan administration.
Close to ten years later the veracity of all sides of the argument
is still subject to debate, and not really germane to the topic
at hand since it is the report itself that is the subject of this
article.

The document's release was announced with a flourish by the
spokesman of the Independent Counsel's office, Charles Bakaly,
who proclaimed: "As required by the Ethics in Government Act,
and with the authorization of the court supervising independent counsels, the Office of Independent Counsel submitted a
referral to the House of Representatives containing substantial
and credible information that may constitute grounds for impeachment of the president of the United States."

The report from Starr included a twenty-five-page introduc-
tory summary, 280 pages of narrative, and about 140 pages dealing with details of alleged impeachable offenses, as well as grand
jury transcripts, videotaped testimony, affidavits, and other materials gathered in the investigation. Starr's office delivered two
sets of the material, or 36 boxes in all.

From a legal standpoint the report had to provide the basis
for the impeachment and removal of the president.

From the political standpoint the report had to provide a
basis that would engender the public's support for such an action.

It failed on both accounts.

On the basis of the evidence provided by the report, the

Republican-dominated Judiciary Committee of the House of Representatives voted out four articles of impeachment (charges against) President Clinton:

1. The President provided perjurious, false and misleading testimony to the grand jury regarding the Paula Jones case and his relationship with Monica Lewinsky.
2. The President provided perjurious, false and misleading testimony in the Jones case in his answers to written questions and in his deposition.
3. The President obstructed justice in an effort to delay, impede, cover up and conceal the existence of evidence related to the Jones case.
4. The President misused and abused his office by making perjurious, false and misleading statements to Congress.

Two of the charges, one count of perjury and one count of obstruction of justice, were voted through by the full House on pretty much a party line vote while the Senate failed to convict the president of these charges, falling short of even a majority vote.

Despite the voluminous materials gathered and the thousands of man hours and millions of dollars spent, the case mounted by the Starr Report failed to pass legal muster in the court of Congress.

But this was not its greatest failure.

The Starr Report was as much a political positioning document as a legal one. It had to sway the general public away from supporting a very popular president by portraying him as having committed offenses that made him unfit for office, not just legally, but morally and ethically as well.

Most people didn't understand such concepts as perjury and obstruction of justice. What they did understand was adultery, lying, cheating, and lying about cheating.

It stood to reason, according to some of the president's opponents, that if the matter was cast in this light, the public would rally against the man in the White House.

As a result when the report was published, the sections included:

II. 1995: Initial Sexual Encounters

A. Overview of Monica Lewinsky's White House Employment
B. First Meetings with the President
C. November 15 Sexual Encounter
D. November 17 Sexual Encounter
E. December 31 Sexual Encounter
F. President's Account of 1995 Relationship

A. January 7 Sexual Encounter
B. January 21 Sexual Encounter
C. February 4 Sexual Encounter and Subsequent Phone Calls
D. President's Day (February 19) Break-up
E. Continuing Contacts
F. March 31 Sexual Encounter

VI. Early 1997: Resumption of Sexual Encounters

1. *Role of Betty Currie*

A. Arranging Meetings
B. Intermediary for Gifts

2. *Observations by Secret Service Officers*

B.   Valentine's Day Advertisement

C.   February 24 Message

D.   February 28 Sexual Encounter

E.   March 29 Sexual Encounter

F.   Continuing Job Efforts

VII. May 1997: Termination of Sexual Relationship

A.   Questions about Ms. Lewinsky's Discretion

B.   May 24: Break-up

These sections (and the supplemental appendix materials) did not skimp on any of the salacious details—to the point that the report itself was banned in certain countries due to the explicit content. A German journalist even tried to sue Starr for publishing pornography on the Internet once the report was posted as an official government document under Starr's authorship.

It was the thinking of Clinton's opponents that the inclusion of the less than prudish materials would attract more readers, thus exposing them to the evidence at hand, and winning them over to the pro-impeachment side.

Attract readers, indeed it did—but that was about it.

In fact, most of the public, having weighed the evidence, seemed to decide that "lying about sex" shouldn't be construed as an impeachable offense, no matter how sordid the details. Just because he might not be the type of guy who you would want to date your sister didn't mean that he was incapable of running the country, and as far as the masses were concerned, his immoral peccadilloes didn't seem to get in the way of doing his job.

Lying about sex was not the same as lying about matters of state, or reasons for getting in a war, or even cheating on your income taxes.

The Starr Report didn't just fail to win most people over to the side of impeachment; it marshaled them against those who did support it. As a result, between the report's publication and the actual impeachment trial, the pro-impeachment Republicans suffered a backlash in the mid-term election and the majority party actually lost almost ten seats in the House of Representatives.

Moreover, President Clinton's job approval rating rebounded and remained high throughout his term, even though his personal approval ratings slipped.

The Starr Report had the exact opposite effect than had been intended. The only architects of its publication who came out ahead were the New York publishers who reprinted it. For them it was just a best-selling dirty book that filled their coffers with profits.

When Good Ideas Are Ignored Just Long Enough to Turn Very, Very Bad

Jaki Demarest

Tax the extremely rich to alleviate the burden on the poor. Sounds like a good idea, right? That was the plan for the Alternative Minimum Tax, soon to be a hot-button issue. If you haven't already heard about it, or worse, been bitten by it, it's all too likely that you will by the end of the decade.

The AMT was initially created to apply to multimillionaires, in an effort to ensure that they paid a more proportional share of taxes. (High-end tax shelters have historically had a tendency to reduce the tax bills of the non-working wealthy to a pittance, with the relatively historically brief exception of the late 1960s and early 1970s, as pointed out by economists Thomas Piketty and Emmanuel Saez in a recent article in the *Journal of Economic Perspectives*.)

The aim of the AMT was unquestionably a good one, but it's beginning to backfire, in part because it's been allowed to ride without being adjusted for inflation for so long, and in part because lawmakers have willfully undermined the tax's intent since 2001, with generous tax breaks for investment income that enable an overwhelming percentage of the richest Americans to escape the AMT altogether.

The AMT was a part of the Tax Reform Act of 1969, and has been in effect since 1970. It was intended at the time to target 155 taxpayer households with high incomes but whom used deductions to pay no taxes in 1966. In the middle of an expensive war in Vietnam, the last thing Congress wanted to hear was that that many multimillionaires were dodging the Federal Income Tax.

In recent years, the AMT has come under increasing fire, but hasn't been either adjusted for inflation or repealed as of the date of this writing, though the newly Democratic Congress has vowed to make an issue of it. Because the AMT has never been adjusted for inflation, an increasing number of middle-income taxpayers have been finding themselves subject to this tax. In the meantime, the multimillionaires for whom the tax was meant have, naturally, found ways to skip out of it entirely, thanks to legislation that favors them.

An estimated 3 percent of those who paid AMT for 2006 are tax-sheltering multimillionaires. The other multimillionaires have found tax shelters that have let them skip out of the AMT altogether. The majority of people actually being caught by the tax are upper middle class, one-quarter of whom are in the $75,000–$200,000 combined family income bracket. (On average, the AMT has boosted those filers' taxes by an estimated $4200 for 2006.)

In point of fact, the AMT can now hit households earning a mere $50,000 annually. Didn't know they were rich, did you?

Under the AMT, capital gains aren't classified as income subject to the tax. And thanks to Bush's tax cuts, the majority of capital gains and dividends are now taxable at a mere 15 percent, down from a top rate of 39.6 percent. Since the majority of income at the top economic tiers comes from capital gains and dividends, as opposed to wages and salaries, the rich and shameless are currently paying taxes at around the same percentage as their underpaid pool boys and golf caddies.

The end result is that the richest taxpayers get a spectacular windfall, while the burden intended for them shifts to others—specifically, the middle to upper-middle class.

By 2007, the AMT will affect an estimated twenty-three million filers. By 2010, an estimated thirty-three million. According to Leonard E. Burman, William G. Gale, Jeffrey Rohaly, and Matthew Hall, better still, the AMT is going to be a bitch to get rid of. An *expensive* bitch—which is why it's been allowed to ride. By 2008, it will cost more to repeal the AMT than to zero out the regular income tax. Without revenue offsets, reforming or revoking the AMT would reduce 2005–2014 revenues by a projected $450 billion if President Bush's tax cuts are allowed to go gentle into that goodnight, and $780 billion if the tax cuts are made permanent.

Simply put, the money to run this country is going to have to come from somewhere, and if the top 1 percent are basking in the sunshine of Bush's tax cuts, it's going to be up to the rest of us to pick up the slack.

It looks fantastic on paper, to 1 percent of us.

"Against stupidity, the gods themselves contend in vain."

—Friedrich von Schiller

Y2K

E. J. Neiburger

One of the strangest phenomena to happen to mankind, especially "civilized" mankind, is the doomsday scenario wherein a careful and well-theorized hypothesis of disaster, documented on paper, grabs hold of society and expands with an uncontrolled life of its own. This seems to happen on the anniversary of every millennium. Thousands of people believed the earth would end the first day of AD 1000. It didn't. And yet the panic returned in AD 2000, commonly called the "Y2K Problem" or just plain "Y2K."

How could this happen? It all started with a simple short cut. When computer memory was dear and every data entry long and hard: why not use only the last two digits of the year? Then came the year 2000, and suddenly there had to be four digits or at least some way to distinguish between 1900 and 2000. Worse yet was what the short entry meant to programs that required the date and simply could not accept one. The result was dire predictions of doom and desolation. But saving those two digits looked like a great idea back when the original computer languages and formats were created. Then those same programmers and engineers, now older and some even wiser, realized what they had done, or what they thought they had done.

We live in an enlightened age of computers, education, free thought, and unlimited communication. We do not believe in devils nor do we burn witches (though in the Congo they burned three hundred people as witches last year). We look at nature with a scientific and objective eye. Yet we had a very odd reaction to Y2K. Our civilization panicked. It went nuts. Dire predictions, repeatedly quoted by the media and supported by many self-proclaimed "experts," spread like wildfire and claimed the ultimate destruction of our civilization. It really looked good on paper. A secret glitch in our most popular computer operating systems (e.g., Windows) would go wild at the stroke of 12:01 after midnight, January 1, 2000, and cause millions of computers to crash, turn off, malfunction, and drag all the data and control systems, depending on those computers, into the depths of non-function, mal-function or worse.

These control systems operate our communications, transportation, banking, industrial machines, weapons—everything. It all seemed possible. People believed it and panicked. Our civilization spent at least six hundred billion dollars to prevent this mythological problem.

Y2K has come and gone. There were a few problems. One army communications system that was operating between posts on computers so old some of them were TRS 80s and Apple 2s went down. A few kids aged a hundred years when their birth certificates were registered on uncorrected sites, but the dire predictions of worldwide shutdowns, mass computer crashes, aircraft falling, missiles detonating, homicidal riots, ill-fitting jock straps, electrical outages, and senior citizens, with starving pets on their laps, dying in their frozen apartments did not materialize. They did not happen in North America, where billions of dollars were spent on computer corrections; they did not happen in Europe where billions of pounds and lire were spent; and they did not happen in countries like Ethiopia, Russia, China, and Gambia, where few funds were made available for their "old" computers. No serious problems occurred at big corporations, which spent millions, or at little one-person dollar stores, which spent nothing for Y2K. Some estimates of the actual money spent on Y2K run as high as 300 billion dollars, when you include the new computers and hardware purchased just to avoid the "problem." It was essentially much ado about nothing.

So much for modern, twenty-first-century man. What first looked like a threat, wasn't. But for all that computer power, no one figured that out. Do you want to take bets on what will happen on Y3K?

Auto Absurdities

Modern Man has an unquestioned love affair with the car. For over a century it has been the focus of the efforts of thousands of inventors and designers. In the nineteenth century doctors were fairly sure that going faster than 50 mph would prove fatal. With the modern auto, superhighways, and the traffic jams we know it is not deadly, just nearly impossible during rush hour.

"We are the first nation in the history of the world to go to the poorhouse in an automobile."

—Will Rogers

Starter Problems

Bill Fawcett

Wheels Within Wheels

By the start of the twentieth century the basic design of a car had pretty well been established. But Milton Reeves just wasn't happy with it. He felt the ride was too rough. Now a lesser man might have invented the wing suspension, but that was a few years off. Mr. Reeves decided the problem was that there were not enough wheels to absorb the shock of that day's rather horrible roadways. So he appeared at the 1911 Indy 500 (the first such race) with his OctoAuto. The eight-wheeled car was long, heavy, and a bit hard to turn, but all that extra rubber gave it a smoother ride. People evidently were not ready to look silly in return for less jostled kidneys and not a single order was made. But not to be discouraged Milton Reeves returned the next year with his Sextauto, which refers to six wheels, not making out in the back seat. But it still turned poorly and cost a lot. Later Mr. Reeves decided to solve a different problem with the early auto and found much greater success. He is the inventor of the muffler.

Something Completely Different

In 1913 the heir to the Scripps publishing fortune decided to make his name by introducing a new and unique auto. He invented the Scripps-Booth Bi-Autogo. He was a good engineer and well funded, so it can only be concluded that it was the desire to create something notably different that led to the introduction of his unique machine. The Bi-Autogo was basically a one-and-a-half-ton motorcycle that held only a driver and a passenger, was powered by the already respected V-8 engine, and steered with a wheel. The vehicle was longer than most cars, heavier, and that V-8 meant it was really fast. Unfortunately it had a tendency to fall sideways at slow speeds, but that was corrected by adding two small wheels that could be lowered by the driver. The big engine proved to be a hit, but not the Bi-Autogo.

No Dealer Add-Ons

More than once the idea of an auto everyone can afford to drive has inspired automakers. In 1920 cars were actually comfortable. Such renowned names as Cadillac, Oldsmobile, and Rolls Royce were producing luxurious vehicles. The engine maker Briggs and Stratton decided they knew how to make a car that everyone could own and drive. The result was the Briggs and Stratton Flyer. The vehicle consisted of a frame, an engine, gas tank, a seat, steering wheel, and four wheels. What it did not include was any body, windows, suspension, lights, or anything else. Basically the Flyer was a frame with a small two-hp Briggs and Stratton engine (yep, the same size a smaller lawnmower has) and a seat. It was in theory a good idea, but slow, cold, uncomfortable, and not very popular.

Fuller Up

One of the least orthodox thinkers of the last century was Buckminster Fuller (1895–1983). The inventor of the geodesic dome excelled at making things happen, but not all of the things he created were good ideas. One of his favorite projects was the Fuller Dymaxion, first shown in 1933. This vehicle, by courtesy a car, was actually supposed to be part of a combination machine that would add jet engines and inflatable wings to allow it to fly. Of course jet engines had not been perfected yet, but that was a tiny detail. So, Fuller created a prototype of the three-wheeled, multi-tone zeppelin on wheels, his first Dymaxion. Seating a dozen and looking more like today's motor homes, the Dymaxion had three wheels with just one in the back that could be turned to any direction. This was to attach to the tail and aileron when the Dymaxion converted into a jet. In the garage it was impressive, or at least impressively large and shiny. On the highway, however, even a moderate breeze threatened the three-wheel design. Two more models were made, the last with a stabilizer fin on the top that gave it an even more "Buck Rogers" appearance. But nothing really helped and the design was abandoned as Fuller went in other directions. But other than being too big, unstable, slow, hard to steer, and expensive, it had seemed like a great idea.

"Soon shall thy arm, unconquered steam! afar / Drag the slow barge, or drive the rapid car; / Or on wide-waving wings expanded bear / The flying chariot through the field of air."

—Erasmus Darwin, 1731–1802

Well, It Worked for Trains . . .

Douglas Niles and Donald Niles, Sr.

The innovations that made the steam engine the driving force of industry were patented by James Watt in 1769. The device went through many improvements during the following century and a half, but the fundamental principle remained the same: fuel was burned in a firebox or furnace attached to a boiler containing water. When the fire got hot enough, the water in the boiler would, well, boil. The resulting steam created pressure, and the pressure of the steam—channeled through airtight, strong pipes and valves—could be used to turn a piston that, in turn, could provide power to an engine or pump.

Steam engines are by nature large, and fairly simple. They are considered "external combustion engines" because the fire ("combustion") does not occur inside the engine itself. The first practical uses of steam engines were made in industry, but soon they had been employed to drive ships, and then trains. Despite their size, the power and capacity of steam engines improved transportation efficiency and reliability, provided that the engine was installed in a device (like a locomotive) that was large enough to carry the engine, along with some other payload.

The more complicated internal combustion engine, wherein the fuel burns right within the engine itself—usually on top of the piston—was a relative latecomer to the history of transportation power. The first was built by Nicolaus Otto in 1876. It was a "four stroke" engine, with each piston cycling through these steps: (1) inlet air and fuel; (2) combustion; (3) expansion; and (4) exhaust. The process was improved upon by Gottlieb Daimler and Karl Benz over the next ten years. (The names Daimler and Benz, of course, remain well known in the auto industry some 120 years later.)

During this same time period, approaching the end of the nineteenth century, the concept of the horseless carriage was coming into vogue. People who had been raised to appreciate the use of trains and steamships were naturally inclined to want to expand this technology toward more convenient individual use, and many inventors strove for ways to install engines in "carriages," thus doing away with the need for—and attendant pollution of—single and double horsepower rides.

Since the steam engine had been around for more than a hundred years, while the gasoline engine was a relatively new-fangled concept, much of the initial design in what would become the automotive industry focused on making steam engines small enough to power individual passenger vehicles. The most notable inventors in this field were the twin brothers, Freelan and Francis Stanley.

By 1897 they had a workable model of a horseless carriage powered by a steam engine, and by 1902 they were ready to commence regular production of their Stanley Steamers. The vehicles were met with enthusiasm, and for a time became a commercial success. Yet, by 1927, the company ceased operations, as the Steamers were so clearly falling behind the smaller, more modern gasoline power autos.

Despite a popular misconception of lumbering locomotives on wheels, Stanley Steamers were sleek, classic-looking vehicles. By 1906, a steamer had established the land speed record in a number of categories, topping out at an eye-popping 127 mph! A number of companies were manufacturing steam automobiles by 1910, and they were engaged in stiff competition, mostly with each other. Yet, 20 years later, they were anachronistic antiques.

Why did the steamer fail to catch on? The reasons lie in the technology itself, in a comparison along which almost all the advantages come down on the side of the gas engine. The internal combustion engine was more efficient: it could travel farther on the amount of fuel carried. It was easier to carry a small supply of gasoline, which would directly fire in the engine, than it was to carry a bunch of coal or firewood that was guaranteed to lose efficiency as it burned to heat up a boiler. The gas engine had more, and more readily available, power—when the operator wanted to increase speed, he could do so quickly, while the steam engine still required heat to build up pressure in that boiler. The final factor was purchase price: A Stanley Steamer would set a person back several thousand dollars, while Henry Ford was able to sell his Model T for only about a tenth of that price.

Here They Go Again . . . the People's Car

Bill Fawcett

The idea is sound. Transportation is fundamental and making a car that everyone can afford grows the economy and also opens the market to a whole new group of auto buyers. Occasionally the idea could work: Look at the VW beetle, made all over the world for almost fifty years. But when this idea goes wrong, it goes very, very wrong.

French Retreat

For a while the French auto industry was known for making sturdy cars. Then along came the Dauphine. This 1956 auto was the final result of the same appealing plan that had tripped up so many other companies: making an inexpensive auto that could be sold to just about everyone. But what seemed, and by the Volkswagen had been shown to be, a winning idea was lost in the execution. The car, and they eventually sold over a million of them, was the Renault Dauphine. To save money the metal in the car was thinned and the paint was so thin that rust began the moment you drove out of the showroom. And it took you a long time to drive away since the underpowered engine took 32 seconds to reach 60 mph. The car accelerated more slowly from a stop than most tractors and many other farm implements. Uncomfortable, barely equipped, and cheaply made, the Dauphine's saving grace was that it was even less expensive than a VW. But it rarely outlasted its own warranty.

Unsafe at Any Speed

The book Corsair *Unsafe at Any Speed* was the final stake in the big car, rear engine auto's checkered career and made Ralph Nader a national figure. The initial idea: if the rear wheels drive the vehicle, why not get rid of the heavy, long transmission and put the engine right over the drive wheels? A perfectly logical objective, but Chevrolet's thinking should have taken a few more steps. The change also brought about a few problems. To start with, all that weight in the back meant less weight in the front and less weight over the wheels that steered the car. This gave the Corvair a tendency to spin out whenever the front wheels lost traction.

There was also the problem of collision performance. Most accidents happen when you are moving forward. Upon impact the front of the car stops moving, but the rear continues to have momentum. With the engine in the back there was a tendency for it to continue moving that brief instant more, putting it into the passenger compartment, to the detriment of anyone who happened to be sitting there—like the driver and passengers.

This problem was already known in Europe. When the Nazi Army occupied Czechoslovakia they confiscated a large number of Tatras, a rear engine Czech auto. Rather quickly these cars were found to be so dangerous that German soldiers were banned from riding in them.

Added to the rest of its safety problems was the fact that the Corvair's single piece steering column had nothing in front of or around it to stop it from being pushed like a pike into the driver's chest in any serious front-end collision. Just to make sure that the relatively light front end made the car a death trap, a gas driven heater was included. This used a gas flame to heat the car rather than the engine's heat, which meant the air in a closed Corvair in winter was a nasty soup of unburned carbons and other gases put out by the heater. Not to mention that the gas in the heater was itself a hazard in any collision. So the Corvair was a good idea, but it was also a deathtrap with several health-threatening design flaws.

Trabant

In 1975, the wise heads of the East German communist state suddenly realized they needed a people's car to compete with the Volkswagen being sold in West Germany. Their answer was the Trabant. Even today you get an embarrassed blush from those few last, hard-core holdout German communists just by mentioning the name. This "people's car" had a body made of fiberglass mixed with plant fibers. The good news was that it did not rust; the bad news was that you could practically put your fist through it. The engine was made to such loose tolerances that it smoked as you drove and generated only feeble acceleration. To save on cost, unnecessary safety items like turn signals and brake lights were never included. Even so, it was all that most people in East Germany could get for many years. Then the wall came down, the Germanys merged, and in the vehicles' last moment of glory the highways running west were filled with "Trabis" packed with everything each family had. Within weeks West Germany was filled with abandoned Trabants whose owners, when given the chance, had discarded them as quickly as they had communism itself.

Communist Redux

The Communist nations were not satisfied with the failure of the Trabant. Yugoslavia too developed a people's car, the Yugo. A list of the 1985 Yugo's failings would simply be that of the Trabant (except for the plastic body) with new additions like constantly stalling engines and failing electrical systems. What was most surprising about this people's auto was that Malcolm Bricklin began importing these beauties to the United States in 1985. The imports featured the same high level of quality as the Yugoslavian vehicles and a few added features such as a rear window defroster. The defroster came in handy as the Yugo engine had a tendency to freeze or even throw a rod or two without any warning. The defroster was said to keep your hands warm while pushing your Yugo off the road. The car provided far more jokes for comedians than sales.

> "The reason American cars don't sell anymore is that they have forgotten how to design the American Dream. What does it matter if you buy a car today or six months from now, because cars are not beautiful? That's why the American auto industry is in trouble: no design, no desire."
>
> —Karl Lagerfeld

Built Ford Tinderbox Tough

Joshua Spivak

In the late 1960s, Detroit's Big Three car manufacturers, General Motors, Ford, and Chrysler, ruled the automotive world with little foreign competition. But cars were becoming more and more expensive. Always looking to increase their market share, Ford Motors Company decided to design a cheap subcompact car especially for frugal shoppers. They created the Ford Pinto, succeeding in the primary goal of keeping the cost of the car down.

Lee Iacocca, then a Ford vice president, trumpeted the Pinto model on the price tag and their corresponding profit margins. Iacocca managed to convince Chairman Henry Ford II that the cheap subcompact would sell. Costing Ford less than two thousand dollars to produce and priced cheaply, the Pinto did indeed outsell many competitors. But the Pinto also had a number of serious design flaws.

First, in order to save valuable trunk space, the gas tank was located in a precarious position behind the rear axle. Second, the Pinto lacked a five dollar and eight cent rubber bladder, which would have contained oil spills when the tank was punctured. Third, a one-dollar plastic baffle to prevent the gas tank from being punctured by sharp bolts in the differential housing was somehow never included in the production model. Finally, the rear bumper offered little more than ornamentation, which further limited the car's ability to handle an impact. As a result, if a car hit the Pinto from behind at speeds over twenty miles per hour, the gas tank would frequently break, fuel would spill, and, often enough, the car caught fire.

Though Ford faced a number of car-fire inspired injury and death lawsuits, the company deftly avoided bad Pinto press. However, a 1972 accident, which caused the death of a woman and severe burns to a thirteen-year-old passenger, led to a highly sensational 1977 trial. As the trial got underway, a Pulitzer Prize–winning article in *Mother Jones Magazine* forever imprinted in the nation's mind the Pinto as a "firetrap" and "the barbecue that seats four." "Pinto Madness," written by Mark Dowie, detailed numerous flaws with the Pinto's design, and claimed that hundreds of people died due to the Pinto's propensity to catch fire. Dowie then launched a devastating indictment against Ford's lack of care for the safety of passengers.

The article claimed there was a simple reason for the automotive giant's failure to include protective designs and devices: the safer engineering was simply not cost-effective. Using cost-benefit analysis memos written by Ford executives, Dowie argued that Ford employed a combination of deliberately specious mathematical models, bureaucratic and legislative delaying tactics, and cold-blooded corporate greed to justify the sale of a dangerous vehicle.

While Ford tried to handle the fallout from the article, the company got slapped with verdicts of five hundred sixty thousand dollars for the woman's family, and two and a half million dollars for the thirteen-year-old boy, in compensatory damages. The jury also saw fit to punish Ford with one hundred and twenty five million dollars in punitive damages (later reduced by the judge to three and a half million dollars).

Despite Ford's denials, the media feeding frenzy increased.

In 1978, a scathing *60 Minutes* report further inflated the casualty figures. The Ford memos on the costs of safety improvements grabbed tons of headlines. Though there have been significant questions as to whether the Pinto was actually as dangerous as its opponents claimed, Ford buckled under the pressure and recalled one and half a million Pintos and thirty thousand Mercury Bobcats (a twin of the Pinto). The recall cost Ford up to forty million dollars. Even worse, Ford faced a 1978 Indiana criminal indictment on charges of reckless homicide. The company managed to beat back the prosecutors, preventing a verdict which would have resulted in fines of up to ten thousand dollars per death, not to mention millions in bad publicity.

But for the Pinto, the damage was done. Sales quickly dropped off, and, after 1980, the car was discontinued. Lee Iacocca, a golden boy of Detroit, also did not escape the Pinto scandal unscathed. He was fired one month after the Pinto recall.

The plan for the Ford Pinto was a solid one. As the Japanese manufacturers would show, Americans wanted cheap cars. They just didn't want them to be tinderboxes on wheels.

Like a fish out of water . . .

The Quirky Little Amphicar

Douglas Niles and Donald Niles, Sr.

It's a classic of James Bond–type spy stories: the snappy sports car that turns into a speedboat, an airplane, perhaps even a submarine. What adventure-loving motorist wouldn't love to be able to alter the medium through which his vehicle traveled? In the early 1960s, a small German company actually made it happen—and it was a design that not only looked good on paper, but actually worked!

It just didn't sell.

The newly designed automobile/boat hybrid was introduced in 1962. Called the Amphicar, it was billed as the first non-military amphibious vehicle ever put into commercial production. Although somewhat boxy by sports car standards, it was a cute enough little convertible, capable of carrying four passengers. Powered by a forty-three horsepower Triumph engine, it could tool along at 70 mph on land. It was equipped with a unique transmission that could deliver power either to the drive wheels, or to the twin propellers located below the rear bumper.

The real fun began when the operator drove his Amphicar down a boat ramp or other sloping incline into the water. The doors were double sealed to make them watertight—the car was equipped with a bilge pump, just in case—and a shift of the transmission lever started the twin props churning. The vehicle was capable of about six knots in the water, and had only about a foot or so of clearance between the water line and the top of the doors—so rough waters were to be avoided! Interestingly enough, the front wheels steered the car both on the ground and in the water, where they created enough of a rudder effect to allow the Amphicar to turn. A second transmission lever allowed the propellers to be reversed for backing up while afloat.

The Amphicar retailed for about $3,300 in 1962—a little less than a Chevy Corvette of the same period. By 1967, the price of the Amphicar had come down to less than $2,000. More than 90 percent of the approximately 4,000 Amphicars produced went to the United States market. The factory, in West Berlin, was geared up for annual production of some 20,000, but in fact never came close to that total—indeed, some parts in the original production inventory never even had to be reordered.

The reasons for the lack of sales success were not due to its faulty design, but perhaps were rooted in the fact that the company employed a full staff of engineers but no marketing or sales department. It was hoped that word of mouth would spread the news about this quirky little vehicle, but that was simply not enough publicity for it ever to really catch on.

Complications also arose in the American market in 1968, when the fledgling Environmental Protection Agency began to introduce emission controls to combat the growing problem of air pollution. The emission restrictions proved too much of an engineering hurdle for the doughty little company, and with the loss of its largest market, it was forced to close.

Still, it was a successful design, and possessed—and continues to possess—a real niche appeal. It has proven quite durable—some 20 percent of them are still around, forty-five years after their introduction—which makes an impressive contrast to the survival rate of heavier and larger cars manufactured during the same period. There remains to this day an International Amphicar Owners Club, with active chapters in Chicago and Minneapolis, among other places. They meet for an annual swim-in at Celina, Ohio. It is also claimed that an Amphicar has crossed the English Channel in bad weather, which is hard to believe for anyone who has seen one afloat. (Perhaps they affixed a very heavy-duty waterproof top!) On the other hand, more than one Amphicar was lost when it entered the water after the owner or operator had forgotten to reinstall the bilge plug.

Perhaps, as with so many inventions, the real success of the Amphicar will be found in the legacy of future amphibious vehicle designs, several of which have been recently introduced. An English company produces the Aqua Sports Amphibian, which is capable of 100 mph on land and some 25 knots in the water—it can even tow a water skier! Although advertised as a commuter vehicle for commuters who wish to avoid traffic jams, the price—a cool quarter of a million dollars—keeps it out of most people's hands.

Another aquatic vehicle, the Hydra Spyder, offers a Corvette engine and a jet drive for water use, and retails for only about $150,000. The Hydra Wind is yet a third amphibious vehicle, which is a combination of a touring bus and a motor yacht, for when you want to take a really large group of friends for a combined water and land ride. On land, it resembles a large motor home, while in the water it resembles, well, a large motor home floating in the water.

Even the original Amphicar is still available, albeit as a collector's item. But a host of Web sites and organizations keep not only the memory, but also the reality, of this cool little car alive, and afloat.

It is safe to say that the world has not heard the last of the amphibious automobile.

Plane Thinking

What goes up must come down. Of course that assumes the plane or jet can take off to start with. From the beginning, the saga of air travel has been punctuated not only by incredible breakthroughs, but also the occasional incredibly bad idea. If you bought this book in an airport and are flying, this editor recommends you maybe skip this section until you are back on the ground.

"If we have learned one thing from the history of invention and discovery, it is that, in the long run—and often in the short one—the most daring prophecies seem laughably conservative."

—Arthur C. Clarke, *The Exploration of Space*, 1951

The Spectacular Failure of the Langley Aerodrome

Douglas Niles and Donald Niles, Sr.

The Man Who Almost Beat the Wright Brothers into the Sky

Samuel Pierpont Langley was the Directing Secretary of the Smithsonian Institution during the latter part of the nineteenth and the early years of the twentieth centuries. Like many other inventors of the period, he was determined to create an engine-powered heavier-than-air flying machine that could carry a man through the skies. He termed this device the Aerodrome, after the Greek for air-runner.

Beginning in 1894, Langley's designs were put to the test. Unlike the Wright Brothers, however, Langley was a project director, not a hands-on technician and inventor. As the director of the largest museum institution in North America, he had many responsibilities to keep him busy, but also had access to a staff of engineers, craftsmen, and technicians who could be assigned to do the actual work of building the flying machine.

Langley's flying experiments were tested from a large boat anchored in the Potomac River near Quantico, Virginia, just south of Washington, D.C. The first unmanned aerodromes failed to fly, but by 1896 Langley's team had created a machine that could be launched by catapult, and was capable of flying more or less straight and level for more than half a mile. Even so, the machine was too small and underpowered to carry a person, so the Smithsonian director turned his efforts toward the creation of an engine powerful enough to lift a heavier machine.

In this he was largely successful, though, as with the rest of "his" work, it was not Langley who did the actual inventing. Furthermore, his estimation of the amount of power required was just that: an educated guess. Unlike the Wright Brothers, who constructed a wind tunnel and studied the behavior of an airfoil under actual conditions of lift—they concluded that about ten horsepower (hp) would be required—Langley's efforts were directed toward building the most powerful engine possible.

The first designer, Stephen Blazer, created a rotary engine capable of generating about eight hp. Next to work on it was Charles Manly, who improved it immensely, and built a truly superior engine. A water-cooled radial design, it was capable of generating some 52 hp.

Although his Smithsonian team had begun working with a four-year head start over the Wright Brothers, in Ohio the Wrights possessed insight and skill that the bureaucratically inclined Langley couldn't match. Based on the success of his unmanned flyer, Langley seemed to think that all he needed was a more powerful engine and a larger machine. He failed to grasp two key details that Wilbur and Orville Wright learned through their own experimentation.

The first of these was the center of balance in an aircraft wing. The director of the Smithsonian understood the principle of the airfoil—the classic wing design that creates low air pressure over the top surface of the wing, and thus results in "lift" from below—but he made a crucial mistake in the application of that principle. Langley decided, logically enough, that the center of balance of an airfoil was down the middle of the wing. Therefore, his wings were supported with guy wires located in that position.

The Wright Brothers determined, through hands-on experimenting, that the actual center of gravity in an airfoil was only about a third of the way back from the leading edge. They actually worked out the equations needed to determine the coefficient of lift, which allowed them to build an engine that was only as powerful as necessary. Furthermore, they understood where the stress would come to bear against the wings. As a result, they placed their supporting wires in that position (something that every model airplane builder understands today). Thus, the Wright Flyer was strong enough to hold together under the stresses of flight.

The second failure of Langley's design is harder to understand: he apparently failed to consider that a flying machine would need to be controlled—i.e., "steered"—through three dimensions. While his aerodrome possessed a crude rudder potentially capable of turning it to the right or left, that rudder was located in the middle of the machine, not at the tail. In addition, it had no wing warping ability at all.

Wing warping, which in the early days literally meant bending the airplane while it was in the air, was a crucial component of the Wright Brothers' design—and, later, a key element of their patent. Perhaps their experience in building bicycles and gliders helped them to grasp this requirement. In any event, the pilot of the Wright Flyer (who lay prone on the lower wing) used his own muscle power to twist and bend the airplane, which was a key component of keeping it in stable flight, not to mention providing at least a limited amount of steering capability.

Aided by the labor of his staff and a War Department budget of $50,000, Langley's Great Aerodrome was completed in October 1903—two months before the Wright Brothers were ready to take to the air. The machine was carried to the same houseboat in the Potomac from which the unmanned flyers had been launched. A courageous volunteer, the aptly named Charles M. Manly, took the controls on October 9. The engine roared, witnesses gathered to observe the historic event, and a catapult snapped forward, propelling the Great Aerodrome into the air.

For about a second. Then the wing twisted out of alignment as the supporting wires failed, and the machine plunged straight down, into the waters of the Potomac. Although Manly was fished out of the water, none the worse for wear, the experiment was a very public failure. Nevertheless, Langley went to work on building another machine, and Manly again volunteered to fly it. This craft was ready by December 8, 1903—which was already five days after the Wright Flyer took to the air at Kitty Hawk, North Carolina—but the date was really rather insignificant as this aerodrome, too, collapsed upon launch and plunged into the river (Manly, who perhaps should have been nicknamed "Lucky," again survived.)

Langley died in 1906, his dream unrealized. The Smithsonian carried on his legacy to the point of having Glenn Curtiss modify the aerodrome in 1914. Curtiss put the wing wires in the proper place, and the machine did in fact take to the air—though it still could not be properly steered or controlled. Even so, this "success" caused the museum's new directors to exhibit the aerodrome prominently, and label it "the first machine capable of manned flight." (Part of the reason for the exaggeration lay in an attempt, ultimately unsuccessful, to void the Wright Brothers' patent on powered aircraft design.)

The resulting display led to a feud with Orville Wright (his brother had died in 1912) that led to the original Kitty Hawk flying machine being displayed in a museum in London, instead of the official American history and technology museum. It was not until 1942 that the Smithsonian corrected the label, explained the details of Curtiss's modifications, and allowed the Wright Brothers invention to claim its true place in the history of aviation.

"There is the right way, the wrong way, and the Navy way."

The First U.S. Navy Catapult Launch

Douglas Niles and Donald Niles, Sr.

The Wright Brothers' successful design of a powered, controllable airplane was tested and improved by the two brothers over the years from 1903 to 1908 under a shroud of secrecy. They were apparently the only designers who understood—probably because of their roots as bicycle makers—that the machine could be steered through turns by banking it to the right or left. They were awarded a patent for their steering mechanism, and took some pains to keep the method secret from competitors.

By 1908, however, they were demonstrating their device for army observers in both the United States and France. While the military uses of flying machines had yet to be demonstrated, forward thinking officers in both countries were gradually beginning to realize that this new invention had a great deal of potential. In the U.S., the initial investment in military aviation came from the Army, with the Navy only slowly taking interest.

However, by the end of 1908, Rear Admiral Cowles, chief of

the Navy's Bureau of Equipment, had seen enough of the new technology to recommend to the Secretary of the Navy that "a number of aeroplanes" should be purchased by the Navy; that these machines should be capable of flying in less than ideal weather conditions, and "be of such design as to permit convenient storage on board ship." Eight months later the request was denied with the explanation that, "The Department does not consider that the development of the aeroplane has progressed sufficiently at this time for use in the Navy."

With this curt dismissal, aviation-minded naval officers were forced to watch as the U.S. Army continued with its experimental flying machines. It wasn't until 1910 that Captain W.I. Chambers, USN, was authorized to learn as much as he could about airplanes from civilian pilots and designers. Although his queries were rejected by the Wright Brothers, he found a willing assistant in one of the brothers' chief rivals, Glenn Curtiss. Chambers arranged for an eighty-three-foot-long platform, sloping downward, to be installed on the prow of the cruiser USS *Birmingham*. On November 14, 1910, pilot Eugene Ely flew a 50 hp Curtiss pusher down that ramp and into the air, for the first successful launching of an aircraft from a ship.

By December of that year, the first United States Navy pilot, Lt. T.G. Ellyson, was authorized to train under Curtiss himself and learn to fly. He was "graduated" after four months of study in San Diego at North Field—which would eventually become the huge San Diego Naval Air Station—with Curtiss reporting to the Secretary of the Navy that Ellyson could operate all Curtiss airplanes, and that he "is a man who will make a success in aviation." By March 1911, Congress provided $25,000 to develop naval aviation, and a few months later the service ordered, from Curtiss, a "hydra-terra-aeroplane" that could fly from land or water and attain a speed of 45 mph. (Of course, since Curtiss lacked access to the patented banking system of the Wright machines, the plane was more difficult to control than those built by Orville and Wilbur Wright.)

Because of the limited space aboard the decks of ships, Curtiss and Ellyson quickly realized the desirability of flinging an airplane into the air with the assistance of a catapult mechanism. They devised a system that relied upon the strength of two men who pulled ropes attached to the airplane's wings, and tested it on a beach in September of 1911. Running the aircraft down an inclined three-wire rigging from a platform 16 feet high, they successfully put the plane into the air, where it flew out over Lake Keuka and landed on the water.

So far, so good. In theory, it should have been possible to do the same thing from a ship. It was not until the next summer, July 1912, that they were ready to try—after making a few improvements. The Navy invested resources in the development of a compressed-air powered catapult system, which was tested on the Santee Dock in Annapolis. The aircraft, with Ellyson at the controls, was attached to the catapult, and the trigger was released.

However, the nose of the airplane was hoisted too far upward, perhaps because of the powerful catapult. In any event, the plane shot up, not out, and was immediately caught in a crosswind that cartwheeled it unceremoniously into the water. Ellyson crawled from the wreckage, soaked and chagrined, but convinced that the problem was solvable.

And it was. By November, they rebuilt the airplane and adjusted the catapult. When Ellyson made his second attempt on November 12, 1912, the plane took to the air. This was the first successful launch of an airplane by catapult. The aviation pioneering team of Curtiss and Ellyson would go on to establish a number of flying firsts, including the development of seaplane operations (with planes landing and taking off on the water) and the use of cranes to hoist and lower amphibious aircraft onto and off of the decks of ships.

World War I was on the horizon and the U.S. Navy was ready to take to the skies.

> **"Airplanes are interesting toys but of no military value."**
>
> —Marshal Ferdinand Foch, in 1911, who became the
> Supreme Commander of Allied forces in World War I, 1918

Where the Buffalo Drones

The Brewster F2A

William Terdoslavich

The Brewster F2A Buffalo sucked. But that would be a mean thing to say. Let's just say it was "developmentally challenged."

At one time in the late 1930s, the F2A took naval aviation from biplanes to the modern age in one very brief swoop.

That was its high point.

And then it was quickly eclipsed as the aviation world kept speeding onward and upward.

Made in Queens . . .

The Brewster Company originally made cars, starting an aviation division in the early 1930s. It was a time when the aviation world was transitioning from the open cockpit biplane, with fabric-covered wings, to the low-wing all-metal monoplane sporting an enclosed cockpit and retractable landing gear.

The Navy was flying the Grumman F3F, a biplane with a squat, barrel-shaped fuselage and an enclosed cockpit. In 1935, it wanted a low-wing all-metal monoplane that could do 300 mph to replace it.

In an attempt to win the Navy contract, a company named Seversky pulled together a version of its P-35 fighter, refitted with a tailhook. Grumman derived its prototype of the F4F Wildcat from the existing F3F. Brewster designed the F2A as a mid-wing all-metal monoplane with flush rivets. Only the flaps were fabric-covered. It was a lackluster design, but wind tunnel testing and a few tweaks brought it up to speed. The Navy gave Brewster the contract in 1938. Make fifty-four airplanes.

Pronto.

Well that was going to be a challenge. The Brewster factory was in Long Island City, Queens—a grungy neighborhood of factories that still had a gritty look long after the manufacturers went away. There was no airfield. The planes had to be trucked to Roosevelt Field (now a shopping mall in Long Island) for final assembly and flight-testing.

It was not until the following June that the Navy finally started getting its F2As. Each plane took flight with an 850 horsepower Wright Cyclone engine. It had a 1,000-mile range. Two machine guns were fitted under the cowling. Two more were placed in the wings.

VF-3, assigned to the U.S.S. *Saratoga*, got the first batch of F2As. By the second half of 1940, the F2A was not living up to its billing. The landing gear proved to be too delicate for the controlled crash landings typically required on a carrier.

On second thought, perhaps the F4F Grumman Wildcat was the better plane.

The Navy dropped Brewster from its lineup and switched to Grumman.

The second disappointment came from Great Britain, then at war with Nazi Germany. The Royal Air Force purchased about 170 F2As, but technical experts faulted the plane for lacking the speed and punch needed to go up against the Messerschmitt BF-109, then the Luftwaffe's first-line fighter. Adding self-sealing fuel tanks and pilot armor only degraded the F2A's performance.

So the British did what the U.S. Navy did—found a way to get rid of the damn thing. In the Navy's case, that was done by giving the F2As to the Marine Corps. For the RAF, the solution was to pack off the Buffaloes, as the British called them, to the Far East.

The Buffalo wasn't good enough to defend London. Perhaps it will be good enough to keep the Japanese away from Rangoon and Singapore?

This was going to be a stretch. The Buffalo suffered from carburetor tuning problems and had trouble reaching 18,000 feet at 295 mph. Climbing to 25,000 feet took a half-hour, not 11 to 12 minutes as promised.

The British believed that their second-rate planes and third-rate pilots were more than a match for the Japanese.

They were proven wrong.

Destroyed in the Pacific . . .

December 7, 1941, would also live as a day of infamy for Great Britain as well as the United States. While Japan attacked Pearl Harbor, it also moved against the Philippines and Malaya in simultaneous operations.

Four squadrons of Buffaloes made up part of the air force defending Singapore. Another Buffalo squadron was posted in

Burma. The Dutch also had a squadron of Buffaloes to help defend the Netherlands East Indies, today called Indonesia. But even ten squadrons would not have been enough, so long as the planes were F2As.

Whether attacked by the Japanese Zero or Oscar fighter plane, the Buffalo was completely outclassed and gunned down very easily. British historian A.D. Harvey sounded a skeptical note in the plane's favor, arguing that the Buffalo did poorly because RAF pilots had little training, compared to Japanese pilots with extensive training and years of experience.

In the end it did not matter much. Japan overran Malaya and captured Singapore. It had no trouble swatting Dutch Buffaloes over the East Indies or RAF Buffaloes over Burma.

Six months later, it was the Marines' turn to take a beating. A squadron of Buffaloes based at Midway Island flew off to defend it from Japanese air attack on June 4, 1942. VMF-221 went up with nineteen Buffaloes—and came back with six. Captain Phillip R. White, who survived the mission, said, "Any commander who orders a pilot out for combat in an F2A should consider the pilot lost before he leaves the ground."

The U.S. Navy had to begin the war with the planes it had, not the planes it wanted to have, to paraphrase Donald Rumsfeld. But the United States does not put up with second best for very long. The Marines and Navy transitioned quickly to the F4F3 Wildcat, while the Buffaloes that continued to come off the convoluted Brewster assembly line went into service as trainers. If it had wings and couldn't fight, some other use could be found for it.

But the first fifty-four Buffaloes the Navy bought never saw service in the Pacific. The Navy was eager to get rid of the Buffalo at any price, so the U.S. State Department found a buyer.

Finland.

Yes, Finland.

Second-rate Plane Destroys Third-rate Enemy

Was the Navy a sucker for buying the F2A or was Finland a bigger sucker to buy them second-hand?

Hard to say, but Finland did come out ahead in the bargain.

In late 1939, the former Russian province was attacked by the Soviet Union. The Finns fought the Russians hard in the Winter War, at first winning the battles, but slowly losing ground over time to the ever-growing Red Army and Red Air Force arrayed against them. Finland needed planes from any nation willing to sell them, and they couldn't afford to be choosy.

Since the Soviet Union was an ally of Nazi Germany at that point, Great Britain and the United States eagerly helped, selling Finland any surplus second- or third-rate aircraft on hand. Finland bought the fifty-four Navy Buffaloes for $54,000 each. The planes were stripped of their arresting gear and life rafts, packed on to steamships, and sent to Norway. From there, they were shipped to the Saab plant at Trollhatten, Sweden, for assembly before being flown on to Finland. Six made it by March 1940, at the tail end of the Winter War. One squadron got the F2As and trained hard to master the beasts.

After the Winter War, Finland, eager to regain land lost to the Russians, became the only democracy to ally itself with Nazi Germany.

To Germany, June 22, 1941, was the beginning of Operation Barbarossa, the invasion of Russia. But to Finland, it was the first day of the Continuation War. The Buffaloes of LeLv

24 were flying against the far more mediocre Red Air Force, wrecked during the Stalin purges of the late 1930s. The Finnish front was not the first worry of the U.S.S.R., which sent its better units to fight the Germans, so the worst of the worst flew against the Finns.

For the Finns it was open season to kill Russians, no bag limit. The Buffalo scored 496 kills with only 19 losses. Thirty-five of Finland's 50 aces scored their kills while flying Buffaloes.

The plane had "redeemed" itself, but it was at the cost of carrying an asterisk in the history books. While the Buffalo sucked, the inexperienced Russian pilots flying the lesser planes of the Soviet arsenal sucked even more.

But that situation was not permanent for the Russians.

The F2A was transitioned out of Finnish service by 1943, outclassed by increasingly more modern Soviet fighters being flown by better-trained and experienced pilots. By the time the Continuation War ended, with Finland again calling it quits, LeLv 24 had given up Buffaloes in favor of Me-109s. Territory regained was lost again to the Russians, and Finland had to learn how to live next door to the Soviets for the entire run of the Cold War.

The last Finnish Buffalo flew in 1948. By that time, the Brewster Company was out of business. Only two Buffaloes remained out of more than 500 made.

Finland has one in a museum.

The United States managed to get its hands on the other surviving Buffalo, salvaged in the late 1990s from the bottom of a Russian lake where it had ditched during World War II. That plane awaits restoration at the National Museum of Naval Aviation in Pensacola, Florida.

A replica of an F2A was also made for the Cradle of Aviation

Museum housed in what is left of Mitchell Field, not far from the Roosevelt Field Shopping Mall.

The museum's F2A placard simply called the Buffalo "controversial."

It's not nice to speak ill of the dead.

(But the Buffalo still sucked.)

> "One of the serious problems in planning the fight against American doctrine, is that the Americans do not read their manuals, nor do they feel any obligation to follow their doctrine."
>
> —*The Soviet Junior Lieutenant's Notebook*, a training manual

Nazi Kamikaze?

The Selbstopfer

Brian M. Thomsen

One of the most memorable sequences in the Stanley Kubrick film *Dr. Strangelove* depicts Major Kong (played by Slim Pickens) riding an atomic bomb, like some kind of metal bronco, to its Soviet target. Though many viewers enjoyed the humor of this "end of the world" scene from the classic cinematic black comedy, most would be appalled to discover how close to depicting reality the scenario actually came.

During World War II the Japanese deployed kamikazes—pilots willing to crash their planes in a final act of destruction to complete a designated mission. Once all of their bombs had been dropped, the pilots of the so-called divine wind were expected to use their plane as a last weapon, aiming it to maximize damage to the enemy on impact, even though this assured their own death.

It was death from the air—a weapon fueled by self-sacrifice.

The Germans had a similar version of death from the air: the unmanned V1 and V2 rockets.

Launched from the French and Dutch coasts and aimed towards England and Scotland, these rocket-propelled bombs were considered reprisal weapons for the Allied bombings of Germany. But despite a well financed and accelerated research program and self-sacrificingly dangerous test piloting by ace test pilot Hanna Reitsch (who risked life and limb to solve certain stability problems in the rocket's frame by literally riding the flying bomb on a test run), the actual targeted results of these attacks caused mostly random damage due to the imprecision of their navigational guidance capabilities.

The problem was simple—precise targeting was not feasible at the distances the rockets had to travel.

As the legend goes, a solution was offered by the infamous Nazi commando Otto Skorzeny. He was no stranger to staring death in the face, having survived numerous missions into enemy territory against seemingly insurmountable odds. On one mission he even landed a glider in the mountains in the middle of a hostile army to rescue Mussolini.

If a pilot-less guidance system was imprecise, he is believed to have suggested, design a rocket that could be piloted until its destination was secured, at which point the pilot could bail out before contact and after the target's destruction was assured.

And thus the plan for the *Selbstopfer* was put on the fast track under the expertise of Ms. Reitsch.

Right from the beginning, however, it was stressed that the German war effort had no intention of launching suicidal kamikaze missions like the Japanese (despite the fact that the English translation of the word *Selbstopfer* is "self-sacrifice"), and an integral part of the planning and design of the new weapon took into account the means of escape for the pilot.

Just under 200 V1s were modified to include cockpits and a steering apparatus, allowing the rocket to be dropped into its flight path from a bomber to be steered by the pilot into enemy airspace and towards the proximity of its target. Once the flying bomb's path was secured, the rocket itself would then be abandoned by the pilot, who would bail out as it made its vertical path downward to explosive contact with the target. To assure all concerned that the steering actually worked, Reitsch once again rode a prototype (unarmed, but properly weighted to assure a test flight consistent with the actual proposed flight; a wooden landing ski was also affixed to its undercarriage) on a test flight to results that were deemed initially adequate to the needs of the planned missions.

Upon approval of the design, close to a hundred pilots were assigned to train for missions that would target Buckingham Palace, Parliament, and other coveted locations calculated to maximize terror and anarchy in the enemy homeland. More strategic locations would also be included, such as bridges in Allied-held Belgium and battlefield command centers that were springing up along the way as the Allies made their moves eastward.

It was during these training runs that the problems began to crop up.

Though the affixed cockpit was fine for Reitsch's slight build, most of the pilots found it to be an extremely tight fit. The resulting leg cramps made the agility required to extricate one's self before impact quite difficult. Moreover, the hatch and latch that affixed the canopy to the cockpit, though it worked fine during the rocket's horizontal flight, often jammed once the vehicle began its vertical descent—the very point at which it was considered locked on to the target—thus greatly impeding the pilot's ability to escape.

Other factors such as the wind resistance caused by the rate of descent, the proximity of the escape canopy to the pulsejet intake, and so forth, all made the real possibility of bailout almost nonexistent—even before taking into account the hazards of parachuting so close to the ground, into enemy territory, and within the blast range of the rocket's impact.

Right from the beginning, no matter how you looked at it, these were going to be one-way missions. Suicide flights, however, were not supposed to be on the agenda.

All claims to the contrary the *Selbstopfer,* despite all of the extra research and development, was no different from the Asian kamikaze. Lacking the Japanese cultural willingness to fall on their sword, the Germans abandoned the project.

> "No weapon has ever settled a moral problem. It can impose a solution but it cannot guarantee it to be a just one."
>
> —Ernest Hemingway

A Little Hard to Swallow

Paul A. Thomsen

Nazi Germany, World War II

During the first hundred years of manned flight, the pursuit of greater speed and the achievement of grander altitudes consumed time, resources, and even human lives to push aviation sciences forward. In the Second World War, both Axis and Allied forces utilized air power in attempts to pound their enemies into submission, but, unlike the other great powers, Nazi Germany was not satisfied with a propeller-driven craft. This desire for something better drove them to make advances in experimental aircraft and engine design. In 1938, the German aviation experts conceived the idea for a line of revolutionary jet aircraft, called the Me-262 *Schwalbe* (German for "Swallow"), a fighter/bomber capable of flying rings around the enemies of the Axis nation. Sadly, for Nazi Germany, no one ever considered their leaders might have even grander plans for the *Schwalbe*, or that problems with bringing a temperamental experimental plane to full battle readiness during the pressures of a losing war might cripple a successful design.

In 1938, Dr. Waldemar Voight and his team conceived of the *Schwalbe* as a fighter plane that would maximize the potential of available technology and revolutionize military aviation. The sleek airframe, held aloft by a 12.8-meter swept wingspan and a turbine engine tucked under each wing, was designed to mirror the speed and dive capabilities of its avian namesake. While a series of design problems plagued the development of the plane's prescribed turbine engines for years, the *Schwalbe* made steady progress from the drawing board to finished product. In fact, the experimental craft's spring 1941 maiden flight, propelled by a set of temporary piston engines, so impressed the German leadership that the designers were asked if it could be refitted for dual use as a fighter and a bomber. Since telling the Führer or his minions "no" was never a bright idea, the designers enthusiastically replied in the affirmative, wiped the cold sweat from their brows, threw out the old design book, and drew up a new one to meet their audience's expectations with what had already been approved.

By 1942, the Junker Jumo 004 B turbine engines were ready to propel the now experimental craft to speeds in excess of 520 mph (more than 100 mph faster than the American-built Mustang fighter), but the *Schwalbe*'s problems were only beginning. First, in trials with the piston engines, the test pilots required lengthier airstrips than other aircraft to be able to achieve enough speed to get the plane airborne. Second, pilots also discovered that just a little too much or too little pitch or yaw on takeoff at the right moment could very rapidly end the life of plane and/or pilot in a blur of speed, metal, tarmac, and trees. Third, once aloft, the pilots encountered increasingly heavy vibrations that violently shook the airframe as the *Schwalbe* reached maximum speed. As a result, while many onlookers marveled at the plane's outward appearing grace and potential combat applications, the pilots and crewmembers struggled to meet the precise demands of the delicate plane.

Once the turbine engines were installed, the test pilots were relieved to find the unnerving vibrations had largely vanished, but while the ride was smoother, the added speed and stability provided by the jet engines added even greater difficulties. The pilots needed even faster reaction times on takeoffs, landings, and, most important, throughout the course of the missions. Before long, it became clear to the leadership that most of the battle-hardened veterans of the Luftwaffe could not meet the demands the *Schwalbe*'s high speed aerial acrobatics, steep dives, and split second attack-or-die decisions made on their bodies and minds. They had to find fresh new pilots with faster reflexes to fly the temperamental little plane.

The Luftwaffe quickly recruited and trained a batch of young aviators. As hoped, after just a few missions, the young pilots rewrote the book on aerial combat. With the fast little plane they could easily overtake their targets and outrun their pursuers. Instead of the old method of attacking from in front or behind their target and then closing in on the enemy, the *Schwalbe* could come at the enemy from the side, fire her four 30mm cannons or, later, rockets, at their target, and fly away before ever approaching the enemy's own effective firing range. In dive bombing attacks, the crews soon discovered the guidance systems were too slow and cumbersome. They learned to rely more on instinct and rapid eye-hand coordination than instrument guidance to effectively drop their bombs and destroy designated targets. By late 1944, the plane was finally fulfilling its designers' dreams.

With this new breed of pilots flying the most advanced set of technological aviation equipment ever designed, what could go wrong?

The design delays, leadership demands, production problems, and the slow learning curve had sorely taxed the nation-state and, with the Allied forces now on the continent, Nazi Germany did not have a lot of time to produce or field the aircraft.

According to the Smithsonian Institute National Air and Space Museum, of the 1,443 Me-262s ever completed, only 300 actually saw combat. With the Allies now bombing as many airstrips as they could find to degrade the Reich's once vaunted air power, pilots and ground crews were forced to hide and fly the battle-ready *Schwalbe* out of unorthodox places, including forested sections of the Autobahn. Likewise, their high-speed engines made turning against incoming enemy fire difficult.

Even more seriously, the *Schwalbe*'s new turbo engines were neither designed nor built for longevity. In the waning months of the war, Nazi Germany's slave labor force manned many of the machines that produced many of the plane's components. Underfed and often abused, these forced laborers did not stress quality control. As a result, the jet turbine engines they built often failed in flight, leaving the startled pilot stranded amid a veritable storm of prey turned predators. Finally, without clear air superiority, the need for lengthy runways made taxiing *Schwalbe* fighters and bombers highly vulnerable to enemy aerial predation.

While as initially problematic as most experimental aircraft, the Me-262 *Schwalbe* actually exceeded her designers' expectations, but neither time nor resources were on her side.

Metaphorically, if only the plane had not been so hard to swallow. . . .

> "Unquestionably, there is progress. The average American now pays out twice as much in taxes as he formerly got in wages."
>
> —H. L. Mencken

Less Bang for More Bucks

The Expensive Saga of the F-111

William Terdoslavich

When you make a new plane, make it high tech. And if that doesn't work, throw money at it until the problems get fixed. That's how the F-111 fighter-bomber got built.

Back in the 1960s, the F-111 was the largest aircraft contract to be put out for bid since World War II. The program was going to run $6 billion, back when $1 billion was a lot of money. The United States planned to buy 1,700 F-111s, then the world's most advanced warplane.

But it didn't quite work out that way.

The F-111's new technology triggered development problems that literally doubled its price before the first plane was ever built.

It got worse from there.

Buying a Paper Airplane

The F-111 suffered its awkward birth because Defense Secretary Robert McNamara made a decision. When faced with two competing bids for a plane to replace the F-4 Phantom II, he picked the most conservative scheme offered by the firm with the most experience building fighter planes. Four boards, staffed by experienced Navy and Air Force officers, chose the Boeing design. McNamara overruled them, choosing General Dynamics. And he made his choice on the basis of blueprints and specifications.

The F-111 program, originally called the Tactical Fighter Experimental (TFX), was going to be a revolutionary break with the past. The plane was designed as a supersonic fighter with twice the range of any existing fighter. While jet fighters of the day could sprint for short distances above Mach 1, the F-111 would fly much of its mission at that speed.

To accomplish this, it would need a "swing wing," which could be moved forward for greater lift during takeoff and landing, but swept back 72.5 degrees to form a delta-shape that would put the wing behind the supersonic shock wave, thus cutting drag. Fuel-efficient fan jet engines would attain the F-111's needed long range, while afterburners would give it extra power for short takeoffs. Advanced electronics would fly it on autopilot below 300 feet. The cockpit design allowed it to eject in one piece, protecting the two-man crew.

In the interests of economy, McNamara wanted the Air Force and Navy to use the same plane, which would help save $1 billion and still keep the F-111's program cost at $6 billion.

But the deal smelled fishy. One senator caught a whiff of it. Then all hell broke lose on Capitol Hill.

McClellan Marches on the Pentagon

Senator John L. McClellan (D-Arkansas), chairman of the Senate Permanent Investigative Subcommittee, was curious about how the F-111 contract had been awarded after Senator Henry "Scoop" Jackson (D-Washington) raised a red flag over the contract. Jackson, of course, was looking out for Boeing, based in his home state.

In Washington, D.C, where everything is political, gossip ran thick that General Dynamics got the TFX contract because it was a Texas-based company. Vice President Lyndon Johnson, Navy Secretary John Connally, and Fort Worth banker Fred Korth, all Texans, were suspected of swaying McNamara's hand.

Boeing chairman William Allen offered to build four TFX prototypes to do a fly-off against four prototypes built by General Dynamics. Boeing came in with a $482 million bid to make 23 prototypes. McNamara suspected Boeing was low-balling the Pentagon.

Roger Lewis, head of General Dynamics, thought that two to four prototypes were too few to test adequately. General Dynamics was also going to build 23 TFX prototypes at $630 million, but that was due to higher labor costs.

McNamara justified picking the more expensive offer by pointing out that General Dynamics, partnered with Grumman as a subcontractor, had more experience making fighter planes than Boeing. The General Dynamics bid relied less on an untested high-tech material called titanium for the plane's design, and an earlier development date was promised.

The hearings got uglier when defense undersecretary Roswell Gilpatric and Navy Secretary Fred Korth (who had replaced Connally) were accused of conflict of interest. General

Dynamics was a client of New York–based corporate law firm Cravath, Swain and Moore, where Gilpatric practiced prior to coming to Washington. Likewise Fort Worth–based General Dynamics had done business with the bank Korth headed prior to his becoming Secretary of the Navy. The Justice Department cleared both men later in 1963, but by then each had resigned from government, going back to the private sector.

Nine months of hearings and 3,000 pages of testimony later, McClellan recessed the hearings. President Kennedy had been assassinated in Dallas.

Lyndon Johnson was now president. General Dynamics was going to make the F-111 fly.

First Flight, Second Guess

Two versions of the F-111 were planned. The Air Force wanted the F-111A to be its supersonic low-level strike aircraft. The Navy wanted the F-111B to do fleet defense, using long-range radar-guided missiles to shoot down hostile aircraft.

The first Air Force F-111A was rolled out in October 1964 and flew by the end of the year. The Air Force wanted to buy 1,400 of the planned 1,700 aircraft. Technical troubles and rising costs now kicked the total program cost estimates $7-$10 billion. The 1,700-plane purchase was cut to 1,450.

The Navy's version went to hell in a hand basket very quickly.

They wanted an aircraft weighing no more than 55,000 pounds. The F-111B was going to weigh in at 68,000 pounds, and it would be twice as expensive as the F-4 Phantom currently in use. By the time the F-111B was rolled out in May 1965, Rear

Admiral William Martin told the Senate that the Navy would rather have more F-4s.

To McNamara, the F-111 was supposed to be the same plane for two different services. It wasn't working out that way.

To do the Air Force's low-level supersonic strike, the F-111's airframe had to be extra rugged. That made it too heavy for the Navy.

As the F-111B prototype went into testing, engineers foresaw its takeoff weight reaching 75,000 pounds—almost the weight of an 18-wheel semi-truck—and it had to be catapulted off of a carrier's flight deck. The F-111B could only deliver 90 minutes on station when the Navy wanted four hours. It was supposed to carry six Phoenix missiles, each one weighing 900 pounds, adding still more weight. And the advanced electronics needed to track 18 targets simultaneously would kick the plane's price tag up to $12 million—six times more expensive than the F-4 and twice as costly as the F-111A.

"Lifting devices" were fitted to the F-111B to overcome the extra weight. They increased the nose-up angle of the plane, making it harder for the pilot to see the carrier flight deck while landing. A weight-cutting program accomplished little. It seemed that every improvement the Navy tried to make only resulted in more delay and expense.

The Navy's F-111B program was out of control. It was three years behind schedule. It was way over budget. And one proto-type had already crashed.

Chief of Naval Operations Admiral Thomas Moorer and Vice Admiral Thomas Connelly, chief of naval aviation, both told Congress in 1967 that they did not want the accursed F-111B. The admirals wanted to take the best of its technology to make a cheaper 52,000-pound fighter that could dogfight. (That later became the F-14 Tomcat.)

The Senate Armed Services Committee happily obliged, voting to kill the F-111B.

Senator McClellan won a minor victory and became the F-111's gravest threat.

Good-bye Nevada, Hello Vietnam

It was now the Air Force's turn to suffer. In March 1968, six F-111s were dispatched to Takhli, Thailand, to do their share bombing North Vietnam. If all went well, the F-111 would replace the F-105 Thunderchief (a.k.a. "the Thud"). The F-111 was limited to flying nighttime strike missions against ground targets in North Vietnam's panhandle region, just north of the DMZ.

The first F-111 was lost within four days of starting operations. Radio contact ceased as the plane was on its way to the target area around Donghoi. While North Vietnam claimed it shot the plane down, the Air Force was more skeptical. Said one officer: "We honestly don't know why the plane is overdue."

A second F-111 was lost two days later over Thailand. Fortunately the crew ejected and survived. The terrain-following radar and its backup failed, automatically pitching the plane upwards to avoid hitting hills or trees, but too low to give the pilot enough altitude to recover from the resulting stall.

Two days later, a third F-111 did not return, like the first, disappearing while en route to North Vietnam. It took a while for the Air Force to find the wreckage. The probable cause of the crash was a tube of sealant left in the plane by a careless worker. The tube froze solid at high altitude and slipped into the plane's flight control mechanism, causing it to jam. Some in the Air Force doubted this was the problem. (No more combat missions were flown. The Thailand deployment ended in November 1968.)

In May 1968, another F-111 was lost on a training flight over Nevada, though the crew managed to eject. Later that month, General Dynamics lost a pre-delivery F-111 at an Armed Forces Day air show in New Mexico. That June, the Air Force grounded all 42 of its F-111As.

By August, Air Force inspectors determined that the failure of a six-inch steel rod in a hydraulic valve actuator was the probable cause of the recent crashes. The valve, located in the F-111's tail, controlled the plane's elevator. The subcontractor had originally used a $100 one-piece high-strength steel rod in the valve, but replaced it in later models with a cheaper $50 rod made from two pieces welded together. The welds snapped, causing the valve to fail—with fatal results.

While this problem was being fixed, another fault came to light in the summer of 1968. An F-111 wing carry-through fitting cracked in a stress test. The carry-through fitting is a 3,000-pound steel wingbox that runs across the belly of the plane, with pivot fittings on either side where each swing wing was attached. If this fitting failed in flight, the F-111 would lose a wing.

At first Air Force inspectors suspected the crack was caused by improper bolting of the carry-through box to the rest of the plane. But that was discounted after 2,500 bolt holes were inspected throughout the F-111 fleet, finding only one flawed hole. The wing boxes were reinforced to prevent future cracking.

The cracking problem reappeared in February 1969, when another stress test produced a crack that was nowhere near the bolt holes on the carry-through fitting. This flaw would limit the F-111's service life to about 1,600 hours of flight time, or about four years of normal flying. The wing carry-through box was supposed to have a 6,000-hour life span, allowing for 10-15 years of flying.

Torture testing began on the wing carry-through box in May 1969, simulating 7.33 Gs of force after the structures spent 24 hours at 40 degrees below zero, followed by extensive ultrasound testing to look for cracks. An $80 million program to fix the wing box problem was drafted for all F-111As then in service. This was later readjusted to $40 million of more modest fixes once the stress problem was better understood.

Other problems became public.

The F-111 was supposed to accelerate from 650 to 1,450 mph in 90 seconds, but took four minutes because it was grossly overweight. One solution was to replace the current engines with the more powerful Pratt & Whitney P-100. But the refit was not possible for later models. This would cost $1 to $2 million per plane, kicking up the F-111's unit cost to the $12–$16 million range

The F-111's jet intakes were located right below the point where the wing joined the fuselage. Turbulent air bleeding off the fuselage was entering the jet intakes, causing the engines to flame out. Tweaking the intake design didn't help much.

While the F-111 was grounded, the Pentagon took the knife to the program again, cutting $1 billion by canceling the purchase of the last 40 planes. Sadly, these were to be the F version of the F-111, fixing many of the under-performance problems that the earlier models suffered.

The plane's deadliest foe (no, not North Vietnam) then took aim at the F-111.

Senator McClellan restarted the hearings he recessed back in 1963. He charged the Pentagon and General Dynamics with hiding the F-111's problems:

The low-level supersonic range of the F-111 was supposed to be 240 miles. But in reality, its "dash range" was only 35 miles. Engineers forecast the metal fatigue problems that were now afflicting the F-111's wing carry-through box which meant that the wings were beginning to fall off during flight. Still the Air Force proceeded with the program despite the warnings, first raised in 1967. Boeing wanted to make the wing carry-through box out of titanium, which is lighter and stronger than steel. McNamara had rejected the idea as too radical, too risky, and not cost effective.

Then it was discovered that General Dynamics withheld information from the 1963 hearings that the F-111 design was 5,000 pounds overweight.

Belatedly, McClellan's committee accused Roswell Gilpatric and Fred Korth of conflict of interest because of prior relationships with General Dynamics, even though the Justice Department had cleared both men of that charge in 1963.

The F-111 contract also had no binding performance guarantees and no penalties for failure to meet performance specifications.

McClellan took his last shot at the F-111 in December 1970, simply by listing the numbers. The U.S. was supposed to get 1,700 F-111s for $6 billion, but instead got only 540 for almost $8 billion, and only the last 100 made came close to meeting performance specifications. By the time the last F-111s were being delivered to the Air Force, the sticker price was hitting $14.9 million.

The pennywise McNamara's pet plane proved to be dollar-foolish, as his planned $1 billion savings turned into several billion dollars of extra expense.

To punctuate the whole mess, another F-111 crashed in May

1971, causing the fleet to be grounded for the sixth time since its service began. The Air Force correctly noted that fewer F-111s were lost in equal amounts of flying time for more well-known aircraft like the F-100, F-101, F-102, F-104, F-105, F-106, and F-4. Then again, all of these aircraft were cheaper than the F-111, which put a real dent in the budget every time one crashed.

"Any time there is anything wrong with the F-111, it's all over the newspapers," said F-111 pilot Lt. Col. Robert Morrison. "But believe me, I've put in plenty of hours on it and the F-111 can do more things than any other aircraft—and it's safe and stable, too."

Pilots who flew the F-111 swore by it.

The critics swore at it.

Vietnam, Libya, Desert Storm

As much as some proponents wanted to use the F-111 to attack its critics, they still needed the plane to bomb North Vietnam.

In the fall of 1972, the Air Force sent forty-eight F-111s to Thailand, hoping the plane's all-weather night-flying capabilities would negate the winter monsoon that would be hovering over Hanoi. By the end of September, F-111 raids were hitting the rail line running from Hanoi to China. Two were lost on operations in a week—Hanoi claimed the planes were shot down. A third disappeared, flying out of Thailand, with no mayday or distress signal. A skittish Pentagon ordered the F-111 grounded again, but it only took a couple of days to check out the planes and restore them to service.

Throughout October and November, another three or four F-111s were lost, but only one was shot down. This was overshadowed by the downing of B-52 bombers. Until then, none had ever been lost, so each shoot down was bad news on page one. The F-111 became a footnote, receiving some relief from negative press.

By 1986, the F-111's problem-plagued reputation was largely forgotten. The U.S. launched retaliatory air strikes against Libya after its terrorists bombed a West German disco, killing one U.S. soldier. Carrier-borne A-6 Intruders and 18 F-111s based in the U.K. struck five Libyan airfields and Libyan dictator Muammar Qaddafi's mansion-like tent. One F-111 was lost for unknown reasons, while another five aborted their missions because of system malfunctions.

By now the F-111s were dropping highly accurate laser-guided bombs, a far cry from the radar-automated bomb drops of the Vietnam War. They wiped out a portion of the Libyan Air Force and damaged airbase runways. Still, there was a minor failure rate. Three bombs hit the French Embassy in Tripoli. Another bomb missed an airfield by two miles and hit a farm, killing 300 chickens. Libyan officials were eager to show off collateral damage, but barred Western journalists from seeing any damage done to military targets.

By Desert Storm in 1990-91, the F-111 drew no more bad press. Only the later E and F models of the F-111 were still flying, but they were so thoroughly upgraded that their all-weather capability was well matched with their accurate delivery of smart bombs with laser targeting. Close to 100 F-111s were deployed. Not a single F-111 was lost in about 5,000 sorties flown. The F-111s were used to deadly effect against hardened Iraqi air defense sites, aircraft shelters, and airfields.

The U.S. finally got its money's worth out of the F-111 before retiring the plane in 1996, replacing it with the near-flawless F-15E Strike Eagle.

Unlike most winged failures, the F-111 did serve long enough to redeem itself. If the U.S. learned anything from the painful F-111 program, it was to break up the most high-tech aircraft contracts into stages. Never again would billions of dollars be spent for a new airplane based on blueprints and specification sheets.

The practice would now be for contractors to fly-off prototypes against each other, with the winner getting the final contract. This way, Uncle Sam would not get the shaft again.

At least we can hope for the best.

A Bone of Contention

The B-1 Bomber

William Terdoslavich

When a plane is a costly failure, don't change the plane. Change the mission. No matter how many times defense critics and anyone with any common sense have tried to ram a stake through the heart of the North American Rockwell B-1 bomber, it keeps coming back to haunt us.

The B-1 was originally designed in the 1970s to be a high altitude supersonic nuclear bomber. An electronic countermeasures (ECM) suite was supposed to foul the electromagnetic spectrum, blinding any Soviet radar-guided surface-to-air (SAM) missile trying to shoot it down. The B-1's swing-wing could be set forward during takeoffs and landings, then pulled back for the joyride past Mach 1.

Then cooler heads prevailed.

President Jimmy Carter and his Secretary of Defense, Harold Brown, axed the program. To many, the hatchet job looked like a typical liberal hit, eliminating a weapons system that could maintain the nuclear balance of terror with the Soviet Union in the expensive, ever-escalating nuclear arms race. Conservative critics fumed that Carter was making America weak in the face of its enemy.

While Carter had many inept moments in his presidency, this was not one of them.

There was no point in buying the B-1, at $200 million a pop, when the B-2 stealth bomber was already in development. A more technologically advanced plane, the B-2 was invisible to radar, thanks to the use of non-reflecting composite materials and its sleek design.

But Carter and Brown could not tell off the critics. The B-2 was still a top secret.

Ronald Reagan to the Rescue!

Not being tough enough on Communism was one of many image problems that Carter failed to shake in his re-election campaign of 1980. Ronald Reagan had no qualms exploiting Carter's weakness in order to win. Reagan kept his promise to revive the B-1, purchasing 100 aircraft for a total program cost reaching $29 billion.

The new plane was rechristened the B-1B, but Air Force personnel nicknamed it "The Bone," for B-one. It was to be the replacement for the aging 1950s-vintage B-52, only this time redesigned as a low-level bomber carrying nuclear weapons. The prototype B-1B Lancer took off in 1983 and entered service in 1985.

But that pre-empted the B-2 stealth program to a degree. Northrop's $25 billion project was more revolutionary than

evolutionary, delivering not a prototype but the first production model during Reagan's presidency. Building out the whole
run of 132 B-2s would have cost about $200 million per plane—
about the same flyaway cost as the B-1. But having both the B-52
and B-1 dampened the need for the B-2. As a result only 18 of
the stealth bombers were purchased. The program's price tag
was spread out over the few aircraft made, raising the sticker
price to $3.2 billion per plane. (Ouch.)

Maybe all this budget grief would have been for naught if
the B-1 had performed as advertised.

It didn't.

The plane could fly. But its ECM suite could not.

The ALQ-161 was supposed to do for the B-1 what AEGIS
did for the United States Navy—provide a means to track
multiple threats in real time and select counter-measures to
neutralize those threats. AEGIS could successfully track over
120 targets in the air, on the surface and beneath the waves.
The more modest ALQ-161 was supposed to protect the B-1 by
detecting and tracking up to 50 simultaneous threats in the air.
But in actual practice, as soon as the 51st threat appeared on
the radar screen, the ALQ-161 shut down. This can be a bad
thing when the sky is full of SAM missiles trying to home in
on your airframe.

In the early 1990s, the Air Force tried to "rephrase the
problem." The ALQ-161 was now meant to handle only the 11
most serious threats under the banner of "Core ECM." Yet the
ALQ-161 couldn't even handle that.

Perhaps more galling was the B-1's non-appearance in the
Gulf War of 1990-91.

To be fair, the plane's proponents argued that the B-1 was
already deployed to handle its nuclear mission, and therefore

unavailable to drop dumb bombs on even dumber Iraqi soldiers occupying Kuwait.

The plane's critics, however, pointed out that the B-1 was also unavailable at least half the time for its nuclear mission. The Air Force wanted a Mission Capability Rate (MCR) of 75 percent. But the B-1's MCR usually came in under that number during peacetime, by about 10 to 20 percent.

The Gulf War was the Air Force's war. And it was never shy about claiming to have won it single-handed. The outcome was going to drive budgets for the next decade.

The START arms control treaty with Russia also called for cuts in the bomber force to accompany the drawdowns in nuclear-tipped ICBMs and SLBMs. The B-1 was going to have to fight for its fair slice of a shrinking budget pie if it was to keep flying.

But how?

When Given a Lemon, Make Lemonade

By 1994, the Air Force decided it was time to reorient the bomber force away from its nuclear mission over to conventional bombing. The B-1 was getting an upgrade program that would allow it to drop dumb and smart bombs. It would need a secure communications package. The Global Positioning System would be added to its navigation suite. And the troublesome ECM system would be upgraded.

"The B-1 has a colorful history," said the polite and understated Lt. Gen. Richard Hawley, the principal deputy assistant to the assistant secretary of the Air Force for acquisition, in a 1994 interview with *Defense Daily*. "We're thus going to have to overcome a great deal of skepticism to persuade people that it can act as the core of our conventional bomber force." Hawley admitted that the B-1's electronic centerpiece, the ALQ-161, was still not living up to expectations after hundreds of millions of dollars spent upgrading the system.

Even after the changes and upgrades MCR rates still sucked, and would also effect the B-1's "cost of ownership." The 28th Bombardment Wing, based at Ellsworth AFB in South Dakota, got by in 1995 with a 55 percent B-1 readiness rate—about 20 points below the Air Force's own benchmark. Within six months, the 28th BW managed to raise its MCR rate to about 65 percent by streamlining maintenance procedures. Having more spare parts available boosted readiness another four points. The 75 percent MCR was attainable for the B-1, but only if there was adequate support infrastructure, maintenance crews and spare parts, all factors that had been diminished by past under-budgeting.

To free up cash for the needed upgrades and properly maintain the bomber fleet, the Air Force decided to mothball 26 B-1s. With over 25 percent fewer planes, more remaining spare parts and maintenance personnel could then be committed to keeping the rest of the B-1s flight ready.

Upgrading Geek Warfare

So if it doesn't do all you want, make it do more. This is called redefining the mission (or saving your stars). The B-1 was to receive a $2.75 billion conventional weapon upgrade to handle the new Joint Strike Weapon (JSOW) and the laser-guided and famously accurate Joint Direct Attack Munition (JDAM) by 2004. The Defensive Upgrade Program (DSUP) divided upgrade into five parts, lettering each set of upgrades from A through E, modifying portions of the ALQ-161 to improve jamming, threat determination, and emitter detection. The ALE-50 towed decoy array would also be added to augment the B-1's limited ECM abilities.

The ALE-50 is a smallish cylinder with tail fins that is trailed out behind the plane on a tether. It emits a radar signal that will show up as a very large blip on an enemy radar screen, or better yet in the radar homing system of an incoming SAM. So long as the emitter is showing blip that is bigger than the plane, the missile is fooled into killing the decoy.

It was not until December 1998 that the B-1 finally proved itself by flying its first combat mission.

Operation Desert Fox was three nights of retaliatory air strikes against Iraq following Saddam Hussein's expulsion of U.N. nuclear weapons inspectors. A pair of B-1s flew out of Oman to hit the Republican Guard barracks at Al Kut. The B-1s were not fitted out yet with the ALE-50 towed decoy and they still lacked the capability to drop Precision Guided Munitions or PGMs. The mission was a success, but the need for an ECM escort was a sad hint that the pre-modified B-1 could not operate in a high-threat environment without help. The Al Kut barracks was located in the heart of the "Super Missile Engagement Zone," a dense cluster of Soviet-made SAMs and AA guns.

The bomber's proponents pointed to the mission as proof that the B-1 could do its job. The critics noted that the plane was finally making its combat debut 12 years after entering service, compared to the six years that elapsed between introduction and combat for the F-117 Stealth fighter (really a strike aircraft).

Another war would only add to the argument.

Operation Allied Force, also known as the Kosovo War,

saw the B-1 fly into action again. Serbia's cultural identity was forged by its defeat at the hands of the Ottoman Turks at Kosovo in 1389, after which most Serbs migrated north. The area now had a majority Albanian population which Serbian strongman Slobodan Milosevic was trying to clear with his army. NATO intervened to prevent the ethnic cleansing campaign from going through. The resulting 78-day war was largely fought by U.S. airpower.

By the spring of 1999, only seven B-1s had undergone the Block D upgrade of the Defensive Upgrade Program (DSUP), which gave those planes the ability to drop smart bombs and operate the ALE-50 towed decoy. Four B-1s (and one spare aircraft) were sent to the RAF base at Fairford, U.K., with another four following later in the air campaign. The aircraft usually operated in pairs, coupled with a flight of B-52s. While the B-1 was JDAM-capable, the scarce smart bombs were saved for B-2 missions.

Again, dumb bombs would be dropped on dumb targets.

The ALE-50 proved its worth. Thirty SAMs were fired at B-1s during the air war, and 10 of those missiles had achieved lock-ons. The towed decoys were able to divert all incoming missiles.

(It should be noted that only one plane was lost during the war—an F-117 Stealth fighter that was downed by a very lucky SAM shot. The wreckage of the plane was promptly airlifted to Russia for technical analysis. Serbian anti-war protestors meanwhile sported signs saying "Sorry, we didn't know the plane was invisible.")

While the handful of B-1s dropped 20 percent of all bombs during Operation Allied Force, Air Force secretary James Roche was critical of the plane's performance. Only Block D aircraft operated during the war, and even then it wasn't until the war's second week, after the bulk of Serbia's air defenses had been suppressed. The Air Force was still trying to make the B-1 operable in a high-threat environment, without success.

Even the plane's readiness rate became an issue—again. During Operation Allied Force, B-1 mission capability rates (MCRs) averaged around 90 percent. Wartime ignores costs in favor of maintaining capabilities, even if that means throwing lots of expensive spare parts at problematic airframes. After the war, MCRs fell back to between 51 and 62 percent—still well below the USAF's requirement of 75 percent mission readiness.

The B-1 saw service again in the Afghanistan War (Operation Enduring Freedom), which followed the Sept. 11, 2001, terrorist attacks.

The eight B-1s used flew only five percent of the missions, dropping 40 percent of the bombs. But Afghanistan had a minimal air force and air defense network, easily destroyed on the first night of fighting. The B-1, equipped with PGMs, now filled in as a "bomb truck," available for on-call air strikes to support U.S. Special Forces and local Northern Alliance militias.

While this helped win the war, it was proof that the B-1 worked best over enemy nations that could not contest its presence. Afghanistan could not even mount a low-level threat to stop the B-1 or any other plane. It was *Star Wars* versus dirt-poor riflemen. The only B-1 that was lost went down in the Indian Ocean due to multiple technical malfunctions. The B-52 also served in Afghanistan as a "bomb truck." And it was a cheaper, older, less troublesome airplane, too.

By the time of the Iraq War in 2003, the B-1 had become even more of a high-tech bomb truck, able to drop JDAMs on demand. Air Force generals raved about the Bone, likening it to a "roving linebacker," as its long range translated into long loitering time over the battle area. The plane flew only one percent of all sorties during Operation Iraqi Freedom, running from March to May in 2003. About 43 percent of all JDAMs dropped were carried by the B-1. But it should be noted that the B-1 flew its missions around 20,000 feet above the battlefield, well above SAM and flak range. The Iraqi air force never put up a fight. Iraq's degraded air defense system was no longer a threat.

By day two of the war, a single B-1 flew over Baghdad, the center of the Super Missile Engagement Zone, albeit with EA-6Bs for jamming escort. The Iraqis fired a salvo of SAMs. None hit. The lone Bone dropped 23 of 24 JDAMs to eliminate six antenna towers that were jamming the Global Positioning System radio signals that U.S. forces relied on for accurate navigation and targeting.

The B-1 finally got its star moment on April 7, when Seek and Destroy got word that Iraqi dictator Saddam Hussein and his sons Uday and Qusay were meeting at a restaurant in Baghdad's Mansour district. It took 47 minutes for the intelligence tip to wend its way through the chain of command to "bombs away." (The B-1 dropped its load 12 minutes after getting the order.) Unfortunately, Saddam and his sons were not having lunch at the restaurant when the bombs hit.

Throughout the war, B-1 crews got targeting information en route or on station. The plane was "on call" with a load of 24 JDAMs, each one able to hit its target accurately. Mission readiness rates were well above 90 percent, as the Air Force threw lots of spare parts and maintenance people at the 10 B-1s deployed for Operation Iraqi Freedom. But the same could be said about the B-52, F-15, F-16, and F-18. Each one was cheaper to own and operate than the B-1, and could drop GPS- or laser-guided bombs just as accurately.

Money, Money, Money

The Air Force did its best to get the most out of its B-1 investment. The plane's awkward status as a "middle child" between the B-52 and B-2 did not help much as the Air Force planned the near future of its bomber fleet.

Northrop Grumman smelled an opportunity, offering to re-start the B-2 line, planning 40 aircraft to be built over 10 years for only $29.5 billion. The offer got no takers at the Pentagon or on Capitol Hill.

Retired Air Force General Richard Hawley testified before Congress that the B-1 and the B-52 should be grounded. "The bomber force ought to have one bomber," Hawley said, while the Air Force was supporting three different planes at greater expense.

In the end, the Air Force did nothing daring. It trimmed the B-1 fleet by a third to concentrate its remaining maintenance dollars to improve readiness.

The upgrade program took the next budget hit. "The tough decision to terminate the DSUP was made because we can really no longer sacrifice capability in pursuit of a post-2010 defense system plagued by escalating problems with cost and schedule," explained Major General John Curley, director of the Air Force global power program. The DSUP suffered three "re-baselinings" and two "cost breaches" from its start in 1997 to its sooner-than-expected demise in 2003.

The cost savings would be used to fund development of Lockheed's AGM-158 Joint Air to Surface Standoff Missile (JASSM).

This weapon would allow the B-1 to hit targets from 200 to 500 nautical miles away. In effect, the B-1 would be assuming the standoff bombing mission that the cheaper B-52 performed for decades.

The technology of air defense is still evolving, but the Bone's ALQ-161 ECM suite will not keep pace. The B-1 will still need an electronic warfare escort to compensate for its ECM shortcomings. Even if the DSUP program were completed, the B-1 would not be able to handle a "non-permissive high-threat environment." That job would still go to the stealthy B-2s and F-117s, or better yet cruise missiles or UAVs.

Another update will be the "sniper pod," which would allow B-1 crews to get a visual on ground targets slated for close air support. During the Iraq and Afghan wars, escorting F-16s had to visually confirm targets before B-1s could paste them with bombs. But the Lockheed "sniper pod" is also slated for installation on the F-15, F-16, A-10, and B-52—all cheaper aircraft that will perform the same mission.

As the B-1 gets older, it will become more expensive to operate, requiring more maintenance hours per hour of flight time. Some day it will have to be replaced, hopefully by a better plane that won't be plagued by budget overruns, a lengthy development time, and balky technology.

Hope is cheap.

The replacement will probably be expensive.

> "We can lick gravity, but paperwork is a bit harder."
>
> —Werner Von Braun

Faster Than a Speeding Bullet

Michelle Poche

What's faster than a speeding bullet and more expensive than the crown jewels? Answer: the super jet called the Concorde. The first civilian aircraft designed for supersonic transport, the Concorde traveled faster than any other civilian aircraft and at more than twice the speed of sound. The jet was widely considered to be one of the greatest engineering accomplishments of the twentieth century. Yet, for all the awe it inspired, it had equal numbers of detractors that reviled the machine which skimmed the boundaries of space and time.

The sleek, sexy jet was expected to be the crowning glory of Anglo-French ingenuity. It could transport passengers across the Atlantic twice as quickly as a Boeing 747 was able to, while serving up gourmet meals, silk pajamas, and travel gifts, all without a cocktail peanut in sight. A flight from London to New York took only three hours and fifteen minutes, instead of the usual seven plus, and was a luxurious experience from start to finish. With Dom Perignon flowing, Maine lobster, caviar, and truffles served on white china dishes, passengers pressed back into plush seats upon takeoff to be whisked to the skies with speeds in excess of 1400 miles per hour. Passengers aboard the Concorde could view the curvature of the earth as they climbed into the stratosphere, a privilege once held exclusively by astronauts or test pilots. And at eleven miles above the earth, the thinner atmosphere revealed the aurora in its full glory. If the departure schedule permitted, one could even watch the sun rise in London and again upon arriving in New York. After such aviation decadence, subsonic commercial airline travel had about as much appeal as a cramped bus ride across the country with the in-laws.

In the mid 1950s, super sonic transport, or SST, seemed poised to be a practical and profitable means of air travel. Developers held dreams of commercial success that would transform passenger aviation and herald a new era of high-speed travel. They believed the increased speed economy would offset the huge amount of fuel needed for supersonic travel. One SST plane could replace three existing planes, creating further savings on maintenance costs. The plane was designed with transatlantic flights in mind, especially the lucrative New York to London route. The developers dreamed of stratospheric profits, but initially, the only thing hitting the stratosphere was the cost of the jet itself.

In 1962, hoping to offset those escalating costs, French and British designers, Bristol and Sud, merged their efforts, in an attempt to co-produce the world's first civilian SST plane. This development triggered panic in the U.S. aviation industry. Fear that this new supersonic transport would outpace the more traditional long-range designs prompted Congress to approve funding for SST development. Congress selected the Lockheed

L-2000 and the Boeing 2707, hoping to engineer an even more advanced aircraft that would be capable of outpacing their European rivals. The promise of national prestige brought the Soviets into this technological competition against the West with their own SST creation, the TU-144. The race was officially on.

The clear leaders of this aviation match-up were the British and French team that secured 200 initial orders for the Concorde, years before it was flight ready. But things got stalled long before takeoff. The environmental movement was on the rise, and as awareness grew, so did concern over the SST's powerful turbo jet engines which produced high levels of nitrogen compounds thought to damage the ozone layer. Additionally, increasing worry over the noise pollution created by the sonic boom from the planes prompted U.S. lawmakers to withdraw funding on SST development, effectively taking the U.S. out of the race for supersonic supremacy.

But the heavily-invested Europeans continued development. By 1975, one by one the world's airlines started withdrawing their options to purchase the Concorde. In the end, only 20 were actually built, and all of them were purchased by Air France and British Airways.

In 1976, those airlines began Concorde flights to the U.S. Although the Concorde seemed to epitomize elegance and luxury for the sophisticated traveler, its glamour could not muffle the noise from the gas-guzzling four engines. The fastest civilian aircraft on earth had earned the distinction of being the loudest plane ever built. On one trip to India, the sound alone caused many windows in the Delhi ATC Tower to shatter. With outcry from residents on both sides of the Atlantic, the Concorde was quickly forbidden from flying supersonically, faster than sound, first over urban areas, and eventually it was restricted to supersonic travel only over the water. Those restrictions caused the web of planned routes to disappear overnight.

In spite of the new restrictions in place, the Concorde continued to be loathed by both environmentalists and local residents in its flight path. It proved to be highly inefficient at low speeds. Merely taxiing down the runway could cost the jet two tons of fuel. One Concorde pilot actually lost his job when he emptied the fuel tanks just taxiing to his gate upon landing. Besides the strikes caused by the noise and expense, the big plane even managed to alienate animal lovers when speculation arose blaming the Concorde for the disappearance of thousands of racing pigeons. A geophysicist in California found evidence that shock waves from supersonic aircrafts destroyed the homing pigeons' ability to hear low-frequency sounds that assisted them on their return flights home.

For years, residents living near New York's JFK airport fought to have noise restrictions placed on the Concorde. While federal laws enacted during that time required other airlines to produce quieter engines, the Concorde was exempt because the engineers could not find a way to make the SST plane any quieter. Even flying sub-sonically, the Concorde and its massive turbo jet engines generated tremendous noise: car alarms screeched, windows shook, and telephone conversation became impossible near the ear-piercing roar of the Concorde taking flight.

With its massive appetite for fuel, its dwindling number of viable routes, and its sparsely populated flights, the Concorde became a financial nightmare.

As if that wasn't bad enough, Air France flight 4590 crashed outside Paris on July 25, 2000, killing everyone on board. The cause? Debris on the runway from a previous, poorly maintained Continental flight shredded a tire. Shrapnel from the tire impacted the Concorde, severing an electrical line and causing a fuel tank to catch fire. The crew tried to compensate by shutting down one engine and using the other to climb out of danger but the damaged landing gear would not retract and the remaining engine surged erratically, causing the plane to pitch, roll, and slam into a nearby hotel. The resulting public relations nightmare heralded the beginning of the end for the already troubled plane. Continental was cited for the poor maintenance that caused pieces to fall from their plane, but it was Concorde and its passengers who paid the price.

Engineers made major improvements to the aircraft, and it made a brief comeback. However, with passenger revenue falling while the costs of maintaining the aging jets continued to rise, it was a losing battle. After the 9/11 attacks drastically decreased air traffic, it became a rout. In 2003 the Concorde was permanently retired. For many, the cost of speed was just too high a price to pay.

"The reason the American Army does so well in wartime, is that war is chaos, and the American Army practices it on a daily basis."

—from a postwar debriefing of a German General

When the Chopper Gets Chopped

The RAH-66 Comanche

William Terdoslavich

Stealthy.
Hard-hitting.
Situationally aware.
Overweight.
Over budget.
Overtime.
Dead.

The RAH-66 Comanche never reached deployment after consuming billions of dollars in its lengthy development. Originally conceived late in the Cold War, the RAH-66 was supposed to help pick out targets for the AH-64 Apache gunship to kill, namely the endless waves of Soviet tanks expected to invade Western Europe.

While the Cold War ended in 1990, the mission of the RAH-66 did not. It was still slated to replace the aging OH-58 Kiowa recon helicopter.

Ironically, the OH-58 outlived its replacement.

Back in the Beginning

The RAH-66 prototype's first flight took place on January 4, 1996. The chopper was a joint effort undertaken by Boeing, which provided the computers and avionic system, and Sikorsky, which made the airframe. The Comanche was supposed to weigh no more than 7,500 pounds, with a flyaway cost of $7.5 million. The Army hoped to buy 5,000 Comanches.

Trouble for the program began early, as the estimated production version of the RAH-66 was 1,000 pounds heavier than its original promised weight. That would mean a slower, shorter-ranged, less maneuverable airframe given its power plant.

The big performance benchmark that the RAH-66 had to clear was the 500-feet-per-minute vertical climb, much needed to avoid small arms fire—something it was not going to do if it was heavier than expected.

Nevertheless, the Army went ahead with the program, setting aside $3.5 billion to complete the next prototype and purchase six early operational versions of the RAH-66. Given the scarcity of budget dollars in the 1990s, many weapons programs were stretching out their development times and going into limited production. But this risked underfunding, which would generate unwanted expenses if you didn't get the program right from the start. There would be little margin for error in the RAH-66 program.

Still, the promised performance benefits were compelling. The RAH-66 had a five-blade rotor, was stealthy, had a fly-by-wire system that substituted a computer for older control cables, a low-drag airframe, and was quiet compared to helicopters of more conventional design. The Hellfire (anti-tank) and Stinger (anti-aircraft) missiles were stored in recessed weapons bays. A three-barrel 20mm cannon would be retractable into the helicopter's fuselage. The radar cross-section of the RAH-66 was only 1/630th of the AH-64 Apache and the infrared signature only one-fifth. The engine, which weighed in at 307 pounds, produced 1,200 horsepower.

Now the trick was to get from first flight to deployment, as the helicopter became more expensive as time passed. The Army's wish was limited to buying 1,292 Comanches, with the first units flying them in 2006.

In a perfect world, this shouldn't be a problem. But real life is never that kind.

In the mid-1990s, an AH-64 adequately performed a recon mission when it checked out Serbian positions in Bosnia during a U.S. peacekeeping intervention. If the AH-64 could do that, then why bother developing the RAH-66?

It was to be the first of many vexing questions.

When Push Comes to Budget Crunch

It seems axiomatic in the Pentagon that the longer a weapons program takes, the more it will cost in the end.

The Comanche program reached $39 billion total. The Congressional Budget office was estimating in 1997 that the helicopter would have a flyaway price of $26 million, well above the $7.5 million originally considered pricey. Upgrading the existing fleet of OH-58s would be a bargain at $1.3 billion over five years.

This took place as the Army was struggling to maintain its 10-division strength in 1996-97. The money needed to do that was being consumed by various weapons programs. Wouldn't it make sense to kill the Comanche, just to free up the money?

The Army answered with a study that concluded that the Comanche had a higher likelihood of survivability compared to the Kiowa.

That's on the battlefield.

But during peacetime, wars are fought over budgets, not terrain.

The Comanche was flying around with a big bulls-eye drawn on it in the shape of a dollar sign.

Development continued through 1997, but the weight problem could not be licked. The Comanche was now up to 8,943 pounds empty—that is, without fuel, ammo, or crew. The budget dollars needed to put the chopper on a diet could not be found. The RAH-66 would go on a development hiatus instead, being grounded for the next 18 months to save money.

Another money-saving move was to push purchase dates back. First it was delayed from 2001 to 2003, then back further to 2006. The longer this program took, the more likely it would be overtaken by events. The General Accounting Office took one look at the need for a recon helicopter and realized it could get the same job done for $4.5 billion if the Army used unmanned reconnaissance vehicles (UAVs). Firepower would be lacking, but the recon capability would be better and the vehicles could be lost without losing aircrew. But the Army was not willing to surrender the program just yet.

The Comanche resumed testing in 1999. Prototype number two took to the air. It was going to be the test bed for the Mission Equipment Package (MEP)—the array of sensors and weapons that would give the Comanche its killing power. Budget questions still dogged the program. There simply was not enough money to both pay for the RAH-66 and keep modernizing existing helicopters.

Modernization was always the cheaper option, especially since it could keep the OH-58 flying until 2025, again undermining the need for the Comanche. In terms of operating costs, a key metric of the peacetime military, the OH-58 only cost $1,100 per flight hour compared to the $2,200 to $2,400 for the AH-64. If the Army purchased the AH-64D upgrade for its AH-64As, it would get the Comanche's capabilities (without stealth) for a much lower cost and still perform the recon mission.

For the RAH-66, the challenge was maintaining justification for an expensive helicopter whose job could be done by cheaper, existing models. The Army didn't see it that way. "We need to get this aircraft (the OH-58) to 2025 and then we have to figure out how we are going to get it out of our fleet quickly," said Brigadier General Robert Armbruster, aviation and missile command deputy for systems acquisition at the Pentagon. "The best thing we can do is to retire it (the OH-58) early and bring the Comanche on board earlier. That is our goal. The second best thing we could do is to minimize the financial impact of the (OH-58) Kiowa Warrior on the aviation budget so that the dollars are available to be spent somewhere else."

In the face of cheaper alternatives that were known to work, the military was choosing the more expensive option.

If You Can't Lose Weight, Add Power

Development woes continued for the RAH-66. The chopper was not coming in at its fighting weight and its performance would be compromised if nothing was done.

A few tweaks were needed, like a slightly longer rotor blade, or changing the performance requirement to 450 feet of vertical climb per minute instead of 500 feet. Engineers looked for ways to get the RAH-66 slimmed down to 8,700 pounds empty. Shedding the synthetic aperture radar and putting it on a UAV was one option. So was ditching the fire control radar, or even opting for a sensor package with fewer capabilities.

Redeveloping the engine offered another option. The Army considered the LHTEC T-801 to replace the T-800 model used on the prototype. The T-801 had 11 percent more horsepower, thus recouping any performance lag caused by greater weight. (LHTEC was a joint venture between Honeywell and Rolls Royce.)

Yet with all these tweaks and fixes, the RAH-66 was still tipping the scales at 9,476 pounds. The Army then tried to offer a performance bonus to the Boeing/Sikorsky team: an extra $5 million bonus if they could get the weight down to 9,300 pounds, an extra $1.4 million if they could get the weight down to 9,250 pounds.

Sadly, they failed.

The nimble wasp they wanted was still more of a fat pig.

And it was putting a strain on the piggy bank.

The longer it took to get the Comanche into production, the more it was going to cost, and it was going to rob funds from existing upgrade programs for the Army's helicopter fleet. Scaling back the OH-58 to 80 percent of what the army wanted might help free up money for the RAH-66. Retiring the OH-58 by 2013 would certainly cut ownership and operating costs. Speeding up the purchase rate of the RAH-66 might also yield another $1–$2 billion to cover the cost of the Comanche program, now creeping up to $40 billion total.

The RAH 66's Engineering, Manufacturing and Development (EMD) stage, which began in April 2000, proved tricky and cost-prone, and was to be completed for $3.1 billion. By December 2001, the EMD was 4 percent over budget and 8 percent behind schedule. Likewise, the Mission Equipment Package development became problematic, as the desired weapons/sensor package triggered a front-end airframe redesign that lagged by four months and the rotor system redesign that came in nine months late.

By December 2002, the Army only had nine more months to work out the weight and integration issues. Even with another engine upgrade to the LHTEC T-802, the RAH-66 was coming in at 9,948 pounds empty—almost one ton above its original design weight. The helicopter was 100 pounds away from maximum design weight and 220 pounds above its maximum empty weight as called for by Army requirements.

The engineers tweaked the RAH-66 some more. Maybe they could find some weight to cut from the rotor system. Perhaps the mission requirements could be limited, so that some equipment could be stripped out of the fuselage. How about carrying two Hellfire missiles instead of four? Strip out the fire control radar? Cut the self-deployment range?

By March 2002, the program was in serious trouble. The EMD portion now required another $3.4 billion on top of the $3.1 billion that was budgeted. The RAH-66 program had already consumed $5.2 billion and the Army only had two prototypes to show for it. Engineers could not add all the computer/sensor features the Army wanted without adding weight.

With money and development woes piling on, Comanche supporters feared the worst: that Congress and/or Secretary of Defense Donald Rumsfeld might notice.

Out Come the Knives . . . and Bullets

The program managers looking at the RAH-66 saw yellow.

The Office of the Secretary of Defense (OSD) saw red. Red is not a good color in politics. It means "danger."

Despite the red flags, the Army thought it could save some money by rearranging the purchase schedule from 72 aircraft a year to 96. It spread out purchase over four distinct blocks of sensor/weapons packages and upgrades. It drew up plans to add capabilities to the earlier block aircraft as new components came on line.

The OSD remained skeptical that the Army could manage speedier purchase of the RAH-66, save money, and still meet the budgetary needs of its current helicopter program and the high-cost Future Combat System (FCS), the Army's biggest budgetary sacred cow.

In October 2002, the Comanche suffered another budget cut among the many that led to its death. Pentagon acquisition chief Peter Aldridge cut the procurement plan from 1,213 Comanches to just 650. This would cut the total cost of the program from $40 billion to $29 billion. The cut seemed pound wise, but was pound foolish, kicking up the price tag to $60 million per helicopter. (The Air Force had fighter jets that were cheaper than this!) And the yearly buy rate was cut from 96 Comanches a year to just 60.

Now the parameters of the RAH-66 program were re-imagined. A yet-to-be designed UAV, that was expected to be out in 2006, would act as an adjunct to the recon copter, feeding it data. But that begged another vexing question: why spend $60 million for a recon copter if a UAV can do the same job for a tenth of the cost?

The next blow to land on the RAH-66 came not from the Pentagon or Congress, but from Iraq.

In 2003, during the initial push on Baghdad, a battalion of AH-64 Apaches had flown a deep penetration raid on Karbala, preparatory to a planned breakthrough there by the Third Mechanized Infantry Division. But the raid went awry as one of the gunships was shot down by small arms fire. The remaining 30 or so Apaches all suffered hits from gunfire and flak. This came within two years of a similar Apache attack being foiled in Afghanistan, again from intense ground fire.

AH-64 tactics underwent revision after Afghanistan and Iraq. Flying at 60 knots was no longer fast enough to evade ground fire. Hovering in place to take aim and fire, a tactic that worked in the Gulf War of 1990-91, was scrapped. Apache pilots now had to execute gun and missile runs while flying forward or diving.

That might extend the usefulness of the existing AH-64, which was already designed with some crew protection. A UAV did not need it since it had no crew to protect. The design emphasis on stealth was now useless for the Comanche. Redesigning the RAH-66 to provide better crew protection, however, would add weight to an airframe that badly needed a diet. And then there was the expendability question. Could the Army afford to lose an RAH-66, at $60 million a pop?

The billions of dollars and years of development time wasted would also draw some unwanted attention from an unlikely source: the new incoming Army chief of staff.

With the departure of Army Chief of Staff General Eric Shinseki in late 2003, the Comanche lost one of its highest-ranking protectors. General Peter Schoomaker, whose background was in Special Forces, was not sentimental about the RAH-66, or any other program that took too long to develop. He was going to take a closer look at Army aviation, whose limited budget dollars became tangled between upgrading existing aircraft and developing the RAH-66.

By this time, the RAH-66 was weighing in at 9,950 pounds and the development team was very eager to trim another 200 pounds. As the engineers thought of ways to cut pounds for millions of dollars, Schoomaker was thinking about cutting programs worth billions of dollars. The RAH-66 was eating up 39 percent of the Army's aviation budget and there still wasn't a production model to show for the effort after a decade of tinkering.

By January 2004, Schoomaker put the Comanche in the crosshairs.

One month later, the general pulled the trigger, killing the RAH-66 program.

"We know it is a big decision and it is the right decision," Schoomaker said.

Acting Army secretary Les Brownlee seconded the opinion: the future operational environment "is inconsistent with the capabilities of the Comanche."

What to Do with the Spare Change?

The Army now found itself with $14 billion in spare change.

Quickly a scheme was drawn up to purchase 368 new reconnaissance helicopters and 303 new light utility helicopters. This would mean turning to cheaper off-the-shelf solutions modified for military use to replace the RAH-66 and OH-58.

The light utility helicopter (LUH) program still looked promising as of December 2006. Sikorsky, Westwind Technology, and EADS teamed up to modify the EC145 Eurocopter for military use. First production models are expected in September 2008, but appear to be delayed having only gotten FAA approval for military use in 2007.

But the Advanced Reconnaissance Helicopter (ARH) being developed by Bell Textron flew into development troubles, just like the RAH-66. Systems integration problems and cost increases are now pushing the price tag per chopper up to $9–$10 million, and the December 2009 delivery date is in danger of slipping. The total program cost of $3.6 billion for 368 airframes grew to $4.7 billion for 512.

And the OH-58 Kiowa?

Its 2013 retirement date got pushed back to 2015, but it is starting to look more like 2017.

The last Kiowa will be older than its pilot by the time it flies into the sunset.

At least it outlived its replacement, the RAH-66 Comanche.

Double Jeopardy

Teresa Patterson

Duplicated systems are usually a good thing, especially in an airplane—except in the case of the Boeing 737.

On March 3, 1991, while on approach to Colorado Springs, United Airlines Flight 585 suddenly veered wildly, turned over, and fell out of the sky, killing all 25 people aboard. The plane involved was a 737. After an extensive 21-month investigation, the National Transportation Safety Board "could not identify conclusive evidence to explain the loss of" the airplane. The board listed the two most likely causes as bad weather or malfunction of the directional control system. The accident joined the short list of unsolved airline accidents, and was largely forgotten—until September 8, 1994, when another 737, USAir Flight 427, suddenly veered wildly and plunged to the ground in a similar uncontrollable dive.

The USAir flight had been on approach to Pittsburgh International Airport, when it encountered turbulence from the wake of another aircraft. Moments after passing through the wake turbulence, the plane veered, turned over, and smashed into the ground, killing all 132 people aboard. Yet the only obvious event, wake turbulence, should have been simply a mild inconvenience, not a catastrophic event. The investigation that followed was the longest aviation accident investigation in NTSB history. There was so little left of the plane that, even after sorting through the rubble, the NTSB could not determine the cause of the crash. The flight data recorder turned out to be little help because it was an older version that only recorded 11 parameters of information. They couldn't blame the weather this time, the day was clear.

A break in the case finally came in 1999, when another 737, this one an Eastwind Airlines flight, barely survived a system malfunction. When the pilot tried to activate the rudder to turn the plane left, he reported that first the pedal pushed back, then suddenly the plane turned right, veering the opposite direction from the controls. Fortunately the Eastwind jet was high enough in the air that the crew had time to react and recover the plane.

Finally with an undamaged plane to study and a live pilot, the NTSB was able to begin testing and determine the probable cause of all three incidents: the redundant systems in the main rudder power control unit. The very redundant systems that were supposed to prevent an accident actually caused it.

The main rudder power control unit controlled the hydraulic fluid flow that activated the movement of the rudder. As all designers know, every crucial system should be built with a backup system—but in the case of the 737's rudder control unit, the backup system was built into the same mechanism as the primary, as a sleeve around the primary, with both valves on the same slide. The NTSB states that the 737 is the only commercial airplane with two wing-mounted engines that was designed with a single-panel rudder controlled by only a single unit that allows the hydraulic fluid to flow into the system to move the rudder, called an actuator. There were also only two valves that controlled the system and both of these dual-concentric servo valves were used jointly instead of two completely separate systems to provide the redundancy needed. All other designs use multiple rudder surfaces and/or multiple rudder actuators. In other words, the backup isn't really a backup—it's still just part of the primary, with a second valve.

While the actuator was designed to allow only one valve to be in position at a time, in actual practice the slide could stick in a neutral position, allowing both valves to activate, causing dual fluid flow and making the rudder to do the exact opposite of what the pilot intended—with the result that a left rudder pedal actually sent the rudder right.

Of course any pilot's instinctive reaction when he says left and his plane goes right is to turn left again, harder—which in this case was exactly the wrong thing to do—causing the rudder to move further in the wrong direction to its "blowdown" or maximum available limit and sending the plane hard over. In the case of Flights 585 and 427, they were too close to the ground when the sudden reversal occurred to have any chance of making corrections—even if they had figured it out.

The NTSB has recommended design changes that would make the secondary system completely separate from the primary, but the FAA says that it will take years to refit all the planes currently in service. Instead of fixing the planes, they offer optional pilot instruction to teach pilots how to react when the unit sticks. In the classes, pilots learn solutions such as releasing the pedal to release the pressure on the unit, or switching the hydraulic master control from primary to secondary for the entire plane to free the rudder.

The NTSB was not impressed. In a 1999 press release, the board stated, "Even with these changes, the 737 series airplanes . . . remain susceptible to rudder system malfunctions that could be catastrophic."

And yet the 737 is still the most widely used jet in commercial aviation. Thousands of people fly in them every day, counting on those optional pilot classes in case of trouble.

Enjoy the friendly skies.

The Plane Was Invisible Until It Was Killed

William Terdoslavich

Remember the A-12 Avenger II?

You don't? Don't worry. You're in good company. Hardly anyone ever knew it was there.

The plane was supposed to be the stealthy replacement for the Navy's A-6 Intruder, a carrier-borne strike aircraft that debuted during the Vietnam War. The A-12 was supposed to be able to carry the AIM-120 AMRAAM air combat missile, the AGM-88 HARM missile to take out SAM radars, an array of bombs, and be able to carry that entire load up to a range of 800 nautical miles at 500 knots.

But we'll never know what the A-12 could do, because according to the Navy they spent close to $5 billion developing the plane *and not one was ever built.*

The A-12 started as a gleam in the eye of the Navy in 1988, when a fixed-cost contract was signed with the development team formed by McDonnell Douglas and General Dynamics. The defense duo was going to design a triangular-shaped flying wing optimized for carrier operations to replace the aging A-6 Intruder and A-7 Corsair II strike aircraft. The Air Force also wanted to get a variant of the A-12 to replace the aging F-111 "Aardvark" and F-15E Strike Eagle. The initial program was going to deliver eight prototype aircraft for just $4.8 billion. Total program cost was going to be somewhere north of $50 billion.

The plane was supposed to be invisible to radar.

It was also invisible to the Pentagon, Congress, and the media for a simple reason: it was top secret.

A portion of that current defense budget was devoted to "black programs," weapons and sensors which, if subjected to open oversight, would surrender a competitive advantage to any peer rival. This same budgetary practice also gave us the B-2 bomber and F-117 Stealth Fighter.

So how bad could this be?

Apparently, things got really bad. And no one knew about it.

By April 1990, then defense secretary Dick Cheney received a Major Aircraft Review of the A-12. It is not publicly known what the review said, but afterwards Cheney trimmed the Pentagon's purchase of A-12s from 858 aircraft down to 620, normally a sign that program costs were becoming excessive. The first flight deadline remained December 1990.

By early May that year, the contractors' development team warned the Pentagon's A-12 program office to expect major delays due to cost problems. This did not seem to discourage the Navy, which went ahead with an option to purchase six production model A-12s for $1.2 billion. This is interesting as not one of the eight A-12 prototypes was finished yet.

The Navy's leadership finally found out that the contractors were having development problems with the A-12. The Navy and the Defense Department each launched their own investigations.

All I Want for Christmas Is My A-12

By November 1990, the murky situation became a little clearer. The A-12 was not coming together as planned and the Pentagon was being kept in the dark. In fact, Pentagon investigators found that the need for secrecy compromised oversight so badly that none could be exercised.

The Navy wanted to have a new airplane by Christmas. But Cheney was in no mood to play Santa Claus. He was getting ready to become Scrooge.

On December 14, 1990, Cheney ordered the Navy to "show cause," explaining why the A-12 should *not* be cancelled. The Navy had three ways it could have handled the situation. It could restructure the A-12 program and stretch it out over a longer period of time. It could give McDonnell Douglas relief from the fixed-price contract. Or it could accept a higher price tag for the airplane, thus helping the contractors recoup their unexpectedly higher costs.

The A-12's costs to date were not encouraging. The program had a $4.8 billion ceiling, but trade press reports put the actual program cost at $7.5 billion, and it was lagging behind schedule by as much as 18 months, pushing the plane's first flight back to 1992. The first production lot of aircraft would be $300 million more expensive than the $1.2 billion price the Navy agreed to.

The Navy had only received $1.2 billion in goods and services to date, and still had no airplane to show for that expense. The Pentagon wanted to recover $1.9 billion in progress payments for work it claimed was unperformed by the contractors.

"No one can tell me how much more it will take to keep the A-12 going," said Cheney. "If we cannot spend the taxpayer's money wisely, we will not spend it." With that said, Cheney took the budget ax to the A-12, chopping the program right out of the budget in January 1991.

The Navy took an ax to its roster as well. Two admirals in charge of the A-12 program were either retired early or transferred, as was a third officer in charge of the program's development. Pentagon purchasing chief John Betti resigned as well, because he had believed the contractor's claims that the program was on track when his own aides told him it was not, according to press reports.

When You Can't Fix a Problem, Fix Blame

Upon terminating the A-12, Cheney declared that the contractors were in default for failing to "design, develop, fabricate, assemble and test the A-12 aircraft within the contract's schedule."

General Dynamics shot back. "This is the result of the government's insistence on a fixed price contract for a major program on the cutting edge of technology," countered company spokesman Alan Spivak. McDonnell Douglas and General Dynamics claimed the program was improperly terminated, adding that the government caused the A-12's costs rise by constantly changing specifications.

Shifting from outrage to chutzpah, General Dynamics and

McDonnell Douglas then sued the Defense Department in 1991.

In late December, 1995, long after the A-12 was buried in a sea of red ink, U.S. Court of Claims judge Robert Hodges found in favor of McDonnell Douglas and General Dynamics. In a 52-page opinion, Judge Hodges declared the contractors were entitled to $1.2 billion to recoup costs suffered from the improper termination of the A-12 program, and could keep the $2.6 billion already paid to them by the Pentagon.

Why Did This Happen?

In doing the autopsy on the A-12, two major factors converged to turn the Avenger into a costly turkey.

The first was the untried technology that had to be integrated into the design, for which no price tag could be affixed before work began.

The second was the fixed price contract that the Navy signed with McDonnell Douglas and General Dynamics. If the cost of developing the plane exceeded the price, then the contractors had to eat the loss. If the project could be brought in below budget, then the profit would be that much sweeter for the contractors.

Weapons systems were usually developed on a "cost plus" basis, where the contractor simply delivered a new plane, tank, or ship meeting the Pentagon's performance specifications. Then they were paid for the weapon.

John Lehman, who served as Navy Secretary during the Reagan Administration, engineered the fixed price contract as a way to save money during an expected dollar drought that would follow the completion of the Reagan defense build-up.

"The moans you hear from the industry in the past few years

are the moans of people facing up to the free enterprise system,"
Lehman told *Aviation Week and Space Technology* in a 1990 in-
terview. "They didn't have a culture of managing costs," he said.
"They manage technology."

But that was not how Eleanor Spector saw it. As deputy assistant secretary for Defense procurement, Spector was the senior civil servant overseeing how our tax dollars were spent to
buy our security. "The overrun on fixed-price contracting is
infinite," she said. "When they [the contractors] start hurting
much, they don't do a good job for us."

Lehman wanted a system where no weapons system went
into production until all problems were worked out. But Spector pointed out that a program's risk cannot be reduced to the
point where there is no further risk.

The A-12 was supposed to be a low risk program.

How did we know that?

General Dynamics and McDonnell Douglas certified it in
writing.

Exit Stage Left

Among the most over-the-top endeavors anyone indulges in is entertainment. The whole idea is to do something so entrancing people will pay to watch or listen to it. The audience demands a steady stream and new and original ideas. So it should be no surprise that some of these ideas turn out to be real stinkers.

"Only thing worse than watching a bad movie is being in one."

—Elvis Presley

Smell-O-Vision

Mixing Odors with Cinema

Douglas Niles and Donald Niles, Sr.

This Movie Stinks! No, Really . . .

The movie industry has always been known for its willingness to innovate and for possessing the creativity to come up with new ideas that seem to stretch the bounds of common sense. Of course, most of the new ideas crash and burn. But, once in a while, a new technique changes the whole industry.

The whole idea of moviemaking as an artistic medium is a good example. It was 1887 when scientist and inventor Eadweard Muybridge set up a number of cameras in a row and used them in sequence to photograph a number of models running past. He was finally able to provide the answer to a question long and passionately debated by equestrians: can a galloping horse have all four feet off the ground at the same time? His images proved it could.

By the late 1890s, Louis Lumière had invented a special camera that could feed a reel of 35mm film past the lens, snapping the shutter thousands of times to take individual pictures. Another invention, the movie projector, could display those images in rapid sequence on a screen, and thus the movie industry was born. The flickering black-and-white images are crude and jerky by today's standards, but a whole industry, including studios and a network of theaters to show the products of those studios, grew up in response to those flickering images. Many theater managers even hired piano players to add a touch of dramatic music to the silent epics projected on the screen.

The cinema flourished during the early decades of the twentieth century, of course, but in 1927 the first of several cataclysmic changes rocked the industry. *The Jazz Singer*, starring Al Jolson, coupled a sound track with a movie, and was an immediate, and colossal, hit. Suddenly, all movies had to have sound!

In the next decade or two, movies introduced images with a barrage of stunning color. Epics such as *Gone With the Wind* embraced the new medium, as did fantasies like the classic Walt Disney animation features. When a dark gray tornado picked up pallid little Judy Garland and deposited her in the domain of *The Wizard of Oz*, the stunning contrast of color's vivid imagery became clear. Suddenly, all movies had to be in color! (Though, primarily because of budgetary concerns, black-and-white film continued to be a common medium into the 1950s.)

Still, those creative minds out in Hollywood were bound to keep thinking, trying to come up with the next big innovation. If we can make movies look and sound spectacular, how about making them *smell* good? Initial attempts included scratch cards, bursts of perfume into the theater, and squirts of aroma at individual seats, but none of these captured the imagination as sound and color had.

But finally, in 1960, one film made a serious attempt to involve audience noses in the movie experience. The film, called *Scent of Mystery*, included some thirty different smells released at each seat in synchronization with the projector. This patented device—surely the inventor was drooling at the prospects of his imminent profits—was called Smell-O-Vision.

Mike Todd, Jr. produced the movie. He, with his father, had created the successful big-screen epic *Around the World in Eighty Days*. For *Scent of Mystery*, he employed the acting talents of Denholm Elliott and Peter Lorre, and even briefly used his stepmother, Elizabeth Taylor, albeit without putting her name in the credits. Bursts of scent would accompany the story as the movie was screened in three specially equipped theatres. A key plot element—a connection between an assassin and pipe smoke—actually relied upon the scent to help tell the story.

Unfortunately, Smell-O-Vision, in practice, simply stank. The odors came out too late or too early, sometimes with overpowering strength, at other times so faint that viewers were forced to sniff loudly as they vainly sought the next clue. The early reviews were terrible, and the movie a complete flop. Mike Todd, Jr. wouldn't be able to produce another movie until he made *The Bell Jar* more than nineteen years later.

And theater-filling Smell-O-Vision would never stink again.

"A real failure does not need an excuse. It is an end in itself."

—Gertrude Stein

Nick & Nora

The Musical

Brian M. Thomsen

Sometimes the potential of the parts do not guarantee a successful whole . . .

Not all Broadway "sure things" are the result of a successful composer/writer/director's next work in the pipeline (as related in the next section on *By Jeeves*).

Sometimes it's the presentation of the perfect package.

An inspired collaboration between playwright and composer under the auspices of a sympathetic and sure director.

A top notch cast with marquee quality names attached that are willing to play up to their potential.

And—perhaps most important—enough time to put the show together right, including an adequate number of workshops and out of town dates to get the production perfectly timed and polished. Put them all together and you have a surefire chance for a successful Broadway run. But on Broadway nothing is ever a sure thing, no matter how good it looks on paper.

Such was the case with *Nick & Nora*.

Before there was *Macmillan and Wife*, *Remington Steele*, or *Moonlighting*, there was Nick and Nora Charles, the archetypal bantering crime-solving upper-crust couple.

The creation of shamus author extraordinaire Dashiell Hammett in his last novel *The Thin Man* (and, according to some critics and friends of the author, based on himself and his paramour Lillian Hellman—though most would agree that Nora was profoundly more likeable than Lillian), Nick and Nora were the idle rich's detectives of choice.

He was a jovial retired Pinkerton (like Hammett himself).

She was a happy-go-lucky heiress (unlike Hellman). Amid a life of parties and the consumption of way too much alcohol, they merrily stumble their way into a crime scene and, over the course of numerous points of befuddlement by the police, wind up solving the crime.

As mentioned before, this was Hammett's last novel, and the only one featuring the Charleses. It is also worth noting that "The Thin Man" of the title does not refer to Nick, but rather to the suspect of whom he finds himself in hot pursuit. Notwithstanding this pesky fact, hordes of fans of the 1934 film version starring William Powell and Myrna Loy (with Skippy the terrier as their beloved schnauzer, Asta) misappropriated the moniker to Powell's character of Nick. That allowed the studios to spawn five additional *Thin Man* movies, a radio series, and a fairly successful two-season TV series starring Peter Lawford and Phyllis Kirk as the sleuthing spouses—and begetting the archetypal TV mystery duo along the way.

Given the elegance of their milieu, the amusement of their banter, and their branded name recognition in the public eye, Nick and Nora Charles seemed tailor-made for the Broadway musical stage.

The transition to stage from book or film is often precarious. It requires a skilled hand familiar with all these modes and acquainted with their virtues and pitfalls. Arthur Laurents was such a man, author of the novels on which his screenplays to the hit movies *The Way We Were* and *The Turning Point* were based, the books for such hit musicals as *West Side Story* and *Gypsy*, and a Tony Award–winning director for the film-based hit musical *La Cage aux Folles*.

Partnered with him on the project was the estimable Charles Strouse, the Tony Award–winning composer of both *Annie* and *Applause*, hit musicals based on high-profile properties (the comic strip *Little Orphan Annie* and the movie *All About Eve* respectively), not to mention his classic toe-tapping contemporary score for *Bye Bye, Birdie*.

There could be no question that these two learned hands understood the tastes of the general public necessary to produce a hit, and with both *La Cage* and *Annie* still very much in the public eye, their names were also worthy of marquee status.

Stepping into the shoes of William Powell and Peter Lawford in the role of the debonair, tippling Nick Charles was Barry Bostwick, who had originated the role of Danny Zucco in the original Broadway production of *Grease*, earned a Best Lead Actor in a Musical Tony for *The Robber Bridegroom*, and had widespread media exposure for his part as Brad in *The Rocky Horror Picture Show* as well as numerous TV mini-series.

Nora was assayed by an equally esteemed Broadway veteran named Joanna Gleason, with numerous Tony nominations to her credit, including a recent win for Stephen Sondheim's *Into the Woods*, as well as prominent parts in a couple of Woody Allen's films.

And some of the names below the titles were no less impressive. Christine Baranski as Tracy Gardner had already been the recipient of two supporting actress Tony Awards and was gradually attracting even more attention to her film and TV work through her turn as Claus Von Bulow's current girlfriend in the movie *Reversal of Fortune*.

Christopher Sarandon played Nick's romantic rival, and though far from a household name (despite a previous Academy Award nomination in the Supporting Actor category for *Dog Day Afternoon*) had managed to maintain a regular profile on Broadway and in numerous showy parts in films such as *Fright Night*. (He also had a profile as Susan Sarandon's first husband, and coincidently married Joanna Gleason after the show closed.)

The unheralded surprise of the show was the talent of Faith Prince in her second Broadway show, after winning a Tony for her debut in *Merrily We Roll Along* two years earlier. Here she was cast in the pivotal role of the murder victim, with flashback sequences featuring her both before and after her demise.

You couldn't ask for a more solid marquee cast.

The show's initial opening date was announced almost a year in advance. This was scheduled to allow plenty of pre-premiere sales to out-of-town theater junkets, as well as set the stage for success through a fall opening that could be promoted through the holiday season (including national coverage via excerpts featured on the Macy's Thanksgiving Day Parade pre-show as well as a float in the parade itself). All along the marketing plan was designed to make the show as much of a hit show for out-oftowners coming to New York on holiday as for the resident New York–area Broadway patrons.

In an era that was growing accustomed to "wait and see" box-office economics, *Nick and Nora* was pre-sold as the guaranteed hot ticket for the year-end holiday season.

With the score and book well in hand, the leads signed, and a production budget befitting its intended success, the schedule was perfectly timed to its overall potential.

Everything was in place for a bankable long-running hit to rival *Annie,* but somehow the show that looked so good on paper did not appear as rosy on the musical stage.

The plot and chemistry strayed from the "Thin Man" formula, a victim of modernization and political correctness. The simple "whodunit" formula may have appeared too dated to modern sensibilities, and gave way to a convoluted *Rashomon* rococo plot that was way too demanding to be taken seriously, and the archetypal Charleses themselves were evolved to a post-*Moonlighting* era of man vs. woman competition (perhaps to replace the now unfashionable tippling that seemed to be so much a part of their characters and indeed their relationship).

And somewhere midway between the time tickets went on sale, and the first scheduled opening night, word of mouth turned ugly—and all of the professionals involved (plus more than a few so-called hired guns) dug in to resolve the problems with changes upon changes upon changes.

Opening night was postponed several times in favor of additional previews to "retool" the show. Over four million dollars in budget overruns were incurred as two weeks of scheduled previews became nine. Several more experts were brought in to observe/comment/consult/fix the show, whose opening date was postponed no less than three times. (This drew a query from the New York City Department of Consumer Affairs, who questioned the suitability of charging patrons who had reserved tickets well in advance for a finished production considering that, since changes were made after almost each performance, the show was obviously still just a work in progress.)

Seventy-one previews passed prior to opening night.

And when opening night finally came, most of the critics were disappointed. It was nowhere near as bad as they were expecting (they had hoped it would unseat *Moose Murders* and *Carrie: The Musical* in the all-time-worst columns)—which is not to say that it was good—or even adequate.

In the words of Frank Rich: "There is no escaping the unfortunate fact that the liveliest thing in *Nick and Nora* is a corpse" (which some PR types tried to spin as a rave for Prince appearing in it, thus implying the show was a "must-see").

The show closed seven performances after its official opening.

The *Times* did a post-mortem after the closing, concluding that some shows should never have been done as musicals but *Nick & Nora* wasn't one of them as "there was considerable potential here—a pair of sophisticated sleuths who have long since laid claim to our affections; a stylish Los Angeles setting at a time when the smog had yet to settle in permanently and the sunsets lived up to the lurid colors in the picture postcards; a murder story, presumably told with a light, bantering touch. Plus Asta, a performing dog."

But all produced naught, as a hit musical that looked good on paper failed to live up to its potential, and the world championship level team that mounted her struck out before the opposing team ever took the field.

> "Everyone judges plays as if they were very easy to write. They don't know that it is hard to write a good play, and twice as hard and tortuous to write a bad one."
>
> —Anton Pavlovich Chekhov

By Jeeves

Brian M. Thomsen

Wunderkind showmen are a curious breed. A string of successes at an early professional age can spawn two resulting situations—an expectation by audiences of something similar of equal or greater quality, as well as, usually at the same time, the desire of the showman to do something new and completely different to show his facility with diverse materials.

This disconnection between audience expectation and "artist's nature" has led to some pleasant surprises—but in most cases, mutual disappointments.

Such was the case with Andrew Lloyd Webber's *By Jeeves*.

Webber, with his collaborator Tim Rice, exploded onto the London/Broadway theatrical scene with two "pop" musicals (then dubbed "rock operas") based on biblical sources. *Joseph and His Amazing Technicolor Dreamcoat* utilized a variety of song styles to tell the Genesis story of Joseph, the son of Jacob

who was sold into slavery in Egypt by his own brothers. It gar-
nered nice notices and helped the authors get backing for their
next project, released as a recording before its stage debut. This
was *Jesus Christ Superstar,* and it took the world by storm in
the early seventies, though some considered it downright blas-
phemous.

First it took Broadway. Then it took London—and the
world.

In no time at all the names Webber and Rice were synonymous with the new performance oeuvre "rock opera," and
Yvonne Ellmann's recordings of "I Don't Know How to Love
Him" and "Everything's Alright" (from *Jesus Christ Superstar*)
were crossover hits that made the show's soundtrack a platinum
seller and laid the groundwork for her later success as a disco
diva with the *Saturday Night Fever* hit "If I Can't Have You."

Needless to say everyone was waiting for the next Webber
and Rice extravaganza.

The only problem was that the two could not agree how to
proceed.

There was a project they were both excited about, but Rice
just didn't feel that it was the right *next* thing. (Rice later admitted his reason for nixing the project was more because he felt
daunted by the source material than by market expectations.)

So Webber, the composer, decided to proceed without him.
The composer of cutting-edge theatrical entertainment daringly
chose to adapt to the musical stage . . . the old-fashioned novels of P.G. Wodehouse—in particularly the stories about Bertie
Wooster and his gentleman's gentleman, Jeeves.

Now Webber was no fool—he realized that he needed a
wordsmith to match his music mastery, and he enlisted the
master playwright Alan Ayckbourn. Ayckbourn's numerous

successes (including *Absurd Person Singular* and *The Norman Conquests* trilogy) had earned him accolades as the Noel Coward of the modern era, making him seem the perfect scribe for the mannered mayhem of the Wodehouse materials.

The *New York Times* was kind in its assessment of the London production, asserting that *Jeeves* was "a well dressed flop." It opened on April 22, 1975, and closed after only thirty-eight performances—and numerous critical drubbings. This was a far cry from the accolades and praise bestowed upon *Jesus Christ Superstar* and *Joseph and the Amazing Technicolor Dreamcoat*. Not only did it fail to become a crowd pleaser and make its way across the Atlantic, it also failed to launch Webber as a bankable commodity sans Rice. (The two reunited for one last collaboration the following year. *Evita*'s success more or less proved that their talents were neither flash in pan nor biblically confined.)

After once again parting ways with Rice, Webber did succeed on his own with such crowd-pleasing, record-setting theatrical blockbusters as *Cats* and *Phantom of the Opera* (and to a lesser extent *Starlight Express* and *Sunset Boulevard*).

Soon success reinvigorated Webber's ego, causing him to look back and question his worst theatrical failure.

The answer was obvious—it wasn't what the audiences were expecting. Nothing wrong with the material—perhaps the timing had been wrong.

Since the now Lord Webber was now the consistent sultan of theatrical success, it was time to revisit the piece. Webber's return to creative status assured backing. A nip and tuck here, drop a song, add a song, plus a bit more dialogue and plot surgery by Ayckbourn, and on May 1, 1996, *Jeeves*, now titled *By Jeeves*, opened in London for a limited season to what one might refer to as "kind reviews"—at least kinder than those of the original production.

Webber interpreted this as the sign that he had gotten it right this time and started plans to slowly move it to Broadway where all of his other shows were flourishing. The fact that the Frye and Laurie BBC versions of the Wodehouse stories had already appeared on U.S. public television had no doubt already prepared American audiences for the sheer delight of the mannerly adventures of a young man and his gentleman's gentleman.

And slowly it did move to Broadway, with all of the hoopla and presale of an in-demand Webber extravaganza from the genius behind *Cats* and *Phantom*.

Despite its being a new version, the *Times* was still unmoved, with the headline on the review "Fooleries From Fey to Tedious," and though it was not panned in the tradition of *Carrie:The Musical* or *Moose Murders*, it failed to secure either accolades or an audience draw beyond the presale.

Perhaps Wodehouse is not the right subject for an American audience who have come to expect pyrotechnics and flash from the Really Useful Lord Webber.

Or maybe it was just not very good Webber.

Its second and probably last incarnation closed after a very short run, having failed to recoup even a substantial fraction of its developmental costs.

> "All you need for a movie is a gun and a girl."
>
> — Jean-Luc Godard

The Big Flop in the Big Top

Brian M. Thomsen

The 1960s saw more than a music revolution in rock and roll; it saw a media revolution as well.

When the Beatles came up with their "innovation" of the "concept album" (or as they referred to it— "an album that could take the place of a tour," the lads from Liverpool having tired of the hassles of the road and the media eye twenty-four/seven) with *Sgt. Pepper's Lonely Hearts Club Band,* which encompassed not just musical evolution for the band, but shades of fictional characterization, theatricality, and elements of plot and mystery (including the still debated "Paul is dead" set of clues) which attracted almost the same amount of attention as a personal appearance without the in-person hassles, it was a win-win scenario.

For their bad boy British rivals, the Rolling Stones, however, it was perceived as the equivalent of a schoolyard challenge. "Keep up or Give up."

So when the Stones' concept album *Their Satanic Majesties*

Request failed to set the world on fire and managed instead to focus more praise on *Sgt. Pepper*, the Stones knew they would have to up the ante even more.

A concept wasn't going to be enough—it had to be a full-out extravaganza. And what better concept for an extravaganza than a circus—a rock and roll circus—a one-night-only themed event, with numerous guest stars leading up to a center ring performance by the Stones themselves.

Allegedly the concept was Mick Jagger's, and he immediately cast himself as the ringmaster. He approached director Michael Lindsay-Hogg to come on board and take the helm so that it could be recorded for broadcast as a BBC event in conjunction with the album release. (Hogg had previously directed the group's performances on numerous UK TV variety shows.)

Immediately an impressive card of supporting acts was lined up, including the soon-to-be ascendant groups The Who and Jethro Tull as well as Taj Mahal and the Stones' woman-in-waiting (a star in her own right) Marianne Faithfull. Also included was a one-time-only super-group called Dirty Mac, which featured John Lennon, Yoko Ono, Eric Clapton, Mitch Mitchell, and Keith Richards on bass guitar.

Jagger and Lindsay-Hogg lined up a sound stage for the performance, while inside they had a seedy big-top circus tent erected, with equally shabby carnival costumes and props supplied for all. They then arranged an invitation-only audience to fill up the bleacher seating and cheer and applaud on cue to increase the circus ambience.

The Stones would take turns introducing the other acts, all leading to a rousing main act of the band performing such already classic Stones hits as "Jumpin' Jack Flash" and "Sympathy for the Devil." Lindsay-Hogg would direct the filming while

Jagger commanded the performances in the center ring. The captured-on-film concert was then slated to be run as a prime time television special on the BBC.

The date of performance was December 11, 1968, and the cameras began to roll at two in the afternoon. Jagger took the center ring in his role as ringmaster, welcomed the audience to the circus, and then introduced Jethro Tull.

Unfortunately, though the man behind the camera was well prepared for the filming, the man in the center ring had basically decided to improvise the entire thing..

Moreover no one had factored in the set-up and take-down time that was necessary between acts, so the show proceeded with many fits and starts at a pace that drained not just the performers but the audience as well. (That there was undoubtedly "much partying" of a sixties sort didn't help matters along; to say nothing of the substance abuse problems of Brian Jones, on the verge of being kicked out of the band due to his drug habit and unreliability.)

As a result the Stones themselves didn't get their performance shot in front of the cameras until many hours later (5 a.m. the following morning), when all parties concerned were on the verge of exhaustion. (All participants credited Jagger's stamina and his embodiment of the ringmaster role as the sole reason that the marathon production made it through to the bitter end.) With the final bars of "Salt of the Earth," the set was struck, and everybody went home.

A rough cut was quickly assembled by Lindsay-Hogg and Jagger, and soon thereafter the BBC airing was cancelled. Unsubstantiated rumors began to fly that the show was cancelled because the Stones had been upstaged by the other bands—but given the conditions all of them performed under that doesn't seem likely; for the most part the performances were all mediocre and not anywhere near the classic levels of rock and roll filmmaking that would soon be attained in *Woodstock* or *Gimme Shelter* or even Lindsay-Hogg's subsequent *Let It Be*.

The bottom line was that they had promised a crowd-pleasing rock'n'roll extravaganza—and had failed to deliver.

The Stones accepted defeat, ended their era of attempting to one-up the Beatles, and placed the filmed footage in a vault, moving on with their careers—including the dismissal of Brian Jones from the band.

The cost of the Rock and Roll Circus was written off as a failed experiment and became the subject of rumors and whispers. With the exception of the soon-to-be-dead-from-a-drug-overdose Jones, all parties recovered from this perceived fiasco (financially if not artistically).

Director Lindsay-Hogg went on to direct the Beatles' final film *Let It Be*, the critically acclaimed BBC mini-series *Brideshead Revisited*, as well as numerous successful London and Broadway productions, such as *Master Harold and the Boys*.

The film remained unreleased until the mid-nineties when, after a few cinema showings, it was made available via videocassette and CD. Most reviews seemed to hail it as more of a landmark curiosity than a classic. A 2004 remastered extended DVD presentation, however, including footage previously thought to be lost as well as behind-the-scenes commentary by Circus participant Pete Townsend, earned much critical praise.

"Its idea of 'production value' is spending a million dollars dressing up a story that any good writer would throw away. Its vision of the rewarding movie is a vehicle for some glamour-puss with two expressions and eighteen changes of costume, or for some male idol of the muddled millions with a permanent hangover, six worn-out acting tricks, the build of a lifeguard, and the mentality of a chicken-strangler."

—Raymond Chandler

The End of RSO

Brian M. Thomsen

Disco Wasn't the Only Thing to Die After *Saturday Night Fever*

If the seventies was the decade of disco then RSO was *the* label of the seventies.

RSO stood for the Robert Stigwood Organization and under its label (which Stigwood formed in collaboration with Polydor) the BeeGees became a supergroup, movie soundtracks became bastions of top-forty hits, and John Travolta became a bankable film star.

Other artists signed to the label included Eric Clapton, Yvonne Elliman, and Andy Gibb, all of whom Stigwood signed before they had broken out as solo artists. But what really put the logo in the public eye was Stigwood's success with movie soundtracks, including such platinum sellers as *Jesus Christ Superstar* (featuring Elliman), *Tommy* (featuring Clapton), and *Fame*, all of which yielded multiple hit singles.

By the late seventies, RSO seemed unstoppable as both the film and soundtrack producer for *Saturday Night Fever*, which set the style for the entire short-lived disco generation, earned John Travolta his first Academy Award nomination, and sold a record number of copies for a double album compilation of disco hits (despite the fact that many of the hits were simultaneously available on competing albums by the artists themselves), including no less than four major hits by the RSO signature group the BeeGees.

Stigwood quickly followed this up with a film version of the long-running Broadway musical *Grease*. This time Travolta (who unbeknownst to his fan base had gotten his start on Broadway in the musical *Over Here*) sang as well as danced, and the show's original score was supplemented with new songs, some sung by oldies stars like Frankie Valli and Frankie Avalon, others by Travolta's film love interest Olivia Newton-John—no stranger to the top forty charts herself. To fill out the mix on the double album, a full side's worth of Sha Na Na hits were added.

Both soundtrack and film were hits of blockbuster proportions.

RSO seemed unstoppable at this point, so Stigwood moved to expand the label's dominance. Having mastered the disco film, and the fifties film, it was time to lay claim to other genres through his innovative style of filmmaking/album marketing.

So over the next two years he worked on two projects to extend his dominance, the first of which he began before *Grease* was yet in the can.

As he had repeated his success with Travolta, Stigwood attempted to do the same thing with the BeeGees though slightly in reverse—where Travolta had to sing the second time, the BeeGees, in their follow-up, had to act.

Their film—the first Beatles film not to feature the Beatles in any capacity—was a "cover" film based on the landmark album of the sixties, *Sgt. Pepper's Lonely Hearts Club Band*. It featured cover versions in character by such hot talents of the seventies as Aerosmith, Earth, Wind and Fire, and Peter Frampton, as well as such oddball vocal inclusions as Donald Pleasance ("I Want You"), George Burns ("Fixing a Hole"), and Steve Martin ("Maxwell's Silver Hammer") as well as, of course, at least an album side's worth of warbling by the BeeGees (the title song, as well as "Getting Better," "She's So Heavy," "She's Leaving Home," "Nowhere Man," etc.).

The plot was at best marginal—a fable of sellout and corruption set against the big and evil business of the music industry (though not billed as such, it could have been the metaphoric memoir of RSO) filled with magic, mayhem, villainy, decadence, and redemption.

Concurrently, though on a slightly slower track since this project was being created from scratch (unlike *Grease* and *Sgt. Pepper*), RSO made a film that would be the metaphoric kid sister of *Saturday Night Fever*, a gritty young outsider's tale that would do for punk and new wave what its older brother did for disco.

This film was called *Times Square*.

Set during the early years of Ed Koch's mayoralty, when New York's fiscal health was on par with the bag lady fashions featured throughout the film, this was the story of a rich girl and a poor girl (literally living out of a shopping cart), both in need of human contact and understanding, who bond amid the squalor and decay of the west end of the Times Square area. It showcased a soundtrack of outré hits by the Ramones, the Talking Heads, Suzi Quatro, Patti Smith, Gary Numan, Garland Jeffries—and, of course, the BeeGees (it was an RSO soundtrack after all). Once again RSO was producing a two-album set that would define the new wave/punk movement as *Saturday Night Fever* had for disco.

Both films followed the RSO formula for success and were expected to yield blockbuster box-office and best-selling soundtracks.

Neither did.

Unlike *Grease*, *Sgt. Pepper* lacked a familiar storyline and believable characters, even within the fantasy world of Hollywood storytelling. Moreover, the madcap mania and mayhem of the previous Beatles films, such as *A Hard Day's Night* and *Help,* didn't work without the Beatles themselves. The powers that be seemed to have substituted "cheesiness" for the "campiness" that worked so well with the fifties homage. Everything looked cheesy—the sets, the costumes, and the performers. Probably the only thing that wasn't cheesy was the budget—and needless to say neither the film nor the soundtrack ever recouped its expenses.

Also unlike Travolta in *Grease*, the BeeGees proved that they were less adept at multi-tasking. They never ventured into the dramatic arts again.

As for *Times Square*, the film did capture some of the neighborhood grittiness that was so evocative in *Saturday Night Fever,* but being down and out in a homeless wonderland on the west side of Manhattan is a far cry from the more familiar neighborhood experiences typified by Tony Manero's Bay Ridge and

Bensonhurst. Moreover, both girls made the audience uncomfortable— as the question of whether they were misfits or real mental cases was purposely left up in the air. There were also several "uncomfortable" sexual subtexts that raised questions and were never resolved. And the music (with the exception of the "let's break out of the Bellevue mental ward" sequence to the tune of "I Want to be Sedated") never seemed to match the loose storyline.

Where *Sgt. Pepper* made the turkey list, *Times Square* just sank into obscurity—no hit records, no box office blockbusters, and plenty of red ink on the RSO balance sheets.

The RSO formula for success had failed, and the company never recovered. In 1981, Stigwood walked away from the company that bore his name and let it be absorbed into the amorphous Polydor label from whence it originally came.

> "Roll the Dice."
>
> —Gary Gygax, co-creator of Dungeons & Dragons

RPG Envy

Brian M. Thomsen

The Plan: Unseat the top role-playing game from its throne and take over the RPG top spot.

The Reality: Since its creation in 1974, the role-playing game *Dungeons and Dragons* has been the top-selling game system of its type. D&D and its progenitor company TSR became the leaders in the field of fantasy role-playing games with numerous rule system innovations and creative and well received "fantasy world settings" such as *Dragonlance*, *Forgotten Realms*, and *Ravenloft.*

Though other companies created other rule systems, and sometimes perhaps edgier settings in which to play, TSR ruled the roost in the game genre they created.

This often left other companies suffering from RPG envy.

One such company, Wizards of the Coast (WOTC), was comprised of "young Turks" whose phenomenally successful collectible card game *Magic: The Gathering* had taken the hobby trade by storm. Peter Adkison, a gnomish systems analyst at

Boeing, founded WOTC in 1990 along with several associates who lucked into success backing the revolutionary card game design work of wunderkind Richard Garfield. The game Garfield had designed was easy to learn, quick to play, and extremely portable—you could play it just about anywhere! And because it was "collectible," players were continually drawn to purchase new packs of cards to supplement their decks and keep themselves competitive with fellow players/collectors. Indeed, the introduction of rare cards and special limited edition cards only succeeded in making the game more "in-demand." The new card booster packs were sometimes dubbed "Crack for gamers."

But the high command at WOTC wasn't satisfied with simply being a successful card game company.

They didn't care that they were the top grossing game company in their field.

They just wanted to be TSR.

No, more than that, they wanted to be a better TSR—and in order to do that they had to design a better role-playing game than *Dungeons and Dragons*. Armed with oodles of cash from the pockets of their card-addicted fans, that is exactly what they set out to do.

Their first attempt was entitled *Everway*.

In an effort to distance themselves from the designosaurs (the game designers working on D&D—or as it was called at the time *Advanced Dungeons & Dragons/AD&D*)—the TSR designers conceived and pitched *Everway* as a form of role-playing for a new generation—dice-less, card-based, and multiculturally mature. Gone were the plastic polyhedrons that determined your fate, to be replaced by a tarot-like Fortune Deck.

Also gone were the hack and slash/spells and monsters, emblematic of the magical medieval setting that had been archetypally associated with *Dungeons and Dragons*. In its place was a pastel multicultural mélange of mythologies carefully selected and designed to offend no open-minded individual and cast off the clichés and stereotypes that had dominated the role-playing realms and fantasies of adolescent boys.

Lead designer Jonathan Tweet and his team delivered a property of which they were justifiably proud. WOTC proceeded to back it with all the marketing clout of its six-hundred-pound pusher of the in-demand *Magic: The Gathering*.

But despite the hype, gamers yawned. *Everway* failed to attract even a substantial fraction of the role-playing audience, let alone compete with *Dungeons and Dragons*. Indeed most D&D players (even the ones who weren't stereotypical adolescent boys) failed to put down their polyhedrons of chance in favor of the new kid on the block. Kinder, gentler, multicultural, and modern just wasn't enough of a selling point for most players; they'd rather hack and slash than switch.

Very quickly WOTC withdrew marketing support from the *Everway* line and adopted a new strategy to achieved domination in the RPG field—if you can't beat 'em, buy 'em out. They decided to buy the company who owned the most successful RPG system and call it their own.

Since TSR had fallen into financial difficulties this became quite easy. WOTC ponied up the finances from their card revenues and bought TSR and all of its brands and trademarks, quickly transferring ownership of all things TSR to Wizards of the Coast.

Now with the hottest card game and the most successful RPG system in their arsenal nothing could stop them—except their inability to leave well enough alone.

No one is ever satisfied with someone else's success, and the

creative instinct that drives gaming culture fosters an evolution
that results in the replacement of the old with the new. Eventually the WOTC design team was no longer satisfied with the inventive and successful playgrounds that had been developed by
the designosaurs at TSR (such as *Dragonlance*, *Forgotten Realms*,
and *Greyhawk*). They immediately launched into a quest for a
new world that they could call their own—one that would succeed and surpass all of these earlier models.

Stage one of this quest was a contest—you propose and design the world and if we pick it we will not just compensate
you but fully develop it and support it across the board with
both games and book products befitting its sure-to-be bestseller
status (which isn't hard to promise when one has no idea how to
make and market a bestseller in the first place).

The winner of the contest was Keith Baker. His wonderfully
innovative world of *Eberron* combined traditional fantasy tropes
with pulp genre elements and quasi-steampunk technology that
begged to be adapted to the big screen by cinematic design ge-
niuses such as Tim Burton or Terry Gilliam. There were elves
and magic but also robots and airships.

Best of all—it was new.

And with the WOTC team firmly behind it, management was
certain the game setting would soon replace the top-selling *Forgotten Realm*'s line. Baker's own novel set in his world would
soon be on the bestseller tables in fine bookstores everywhere.

But as with *Everway*, it just didn't happen that way. *Eberron*
didn't even rattle the *Forgotten Realms* throne. WOTC had no
success with creating or launching new products to replace old
ones.

If something had a mature following already, they could
maintain it (like R.A. Salvatore's best selling *Drizzt* novels, or

the *Forgotten Realms* role-playing setting that had been created by Ed Greenwood), but when it came to making and selling something out of whole cloth, they couldn't make it happen.

The lightning that WOTC caught in the bottle with *Magic: The Gathering* was destined never to return—which turned out to be okay since the company made oodles more dollars with a licensed card game based on the animated series *Pokemon*. If *Magic* was crack for gamers, *Pokemon* was crack for gaming kids.

It made tons of money. But no one was ever going to mistake it for a fantasy role-playing game. As a result, WOTC will always be known first as a card game company and only second for its acquired association with *Dungeons and Dragons*—the role-playing system that still reigns supreme.

"Football combines the two worst things about America: it is violence punctuated by committee meetings."

—George F. Will

The XFL

Brian M. Thomsen

Baseball may be America's favorite pastime, but football or, more precisely, NFL football, is "must see TV."

Not only is the season more limited (only sixteen games), the schedule is almost entirely confined to Sundays with most broadcasts limited to local areas or pay TV venues—and when it comes to TV ratings, it kicks the behinds of all other sports events in the United States.

The problem is simple—it dominates only one third of the year.

Logic has it that if a palatable football substitute were offered in another season, respectable ratings should be assured. But such was not the case during trial broadcasts of such reasonable substitutes as the CFL (Canadian Football League), NFL Europe, and Arena Football, all of which performed on par with American soccer—which is to say, they were not worth the major networks' valuable airtime.

Since a reasonable substitute failed to make the grade, perhaps an "unreasonable" substitute might do the trick.

Thus the XFL was born.

Citing the ratings success that pro-wrestling evolved into once it became more outrageous and less rules conscious, NBC and WWF (World Wrestling Federation) formed an alliance to spawn an equivalent "smash mouth" version of the NFL for broadcast in the television season immediately following the Super Bowl.

Embracing the two themes of sex and violence, the XFL ("X-treme Football League") was sold as football without the drawbacks of civilizing rules—ergo more roughness, fewer penalties, and, in general, more anarchy on the field. It would feature "tramped up" cheerleaders, encourage gutter talk, and embrace "bad boy" branding. Even the team names were designed to invoke a sense of criminality—the Orlando Rage, the NY/NJ Hitmen, the Chicago Enforcers, the Memphis Maniax, and so forth.

Spokespeople emphasized that all points in the game had to be earned through combat (which really only meant that the field goal extra point after a touchdown was eliminated in favor of another play action down at the two-yard line—no different than the NFL's "two point conversion," though the XFL's down only earned one point). Representatives bandied phrases like "mortal combat," "gladiatorial combat," and "to the death" to promote the sense of savagery and danger soon to be on display.

Just as with television's latest incarnation of pro-wrestling, the key concept was to deliver a fan-friendly product that would produce high ratings by providing the audience with the type of sports entertainment they enjoyed magnified to the extreme.

The league signed players, held practices, designed cheer-leader uniforms, and with the signing of Jesse Ventura as the official XFL announcer, brought everything to readiness for the league's big debut.

Opening night, February 3, 2001, more than fourteen million viewers tuned in to watch the Hitmen take on the Outlaws with ratings that dwarfed that season's NFL's Pro Bowl telecast—a debut to be envied by any network.

The problem was that viewers did indeed tune in, but they weren't particularly impressed.

The actual rules variations from the NFL seemed arbitrary and didn't really affect the game play that much (with the possible exception of the man on man conflict for possession of the ball that replaced the opening half coin toss—and also resulted in the first player injury of the season). Moreover the skill of the players themselves was far inferior to the level that viewers of the NFL had come to expect.

And what of the promise of sex and violence? Well, that had definitely been oversold—it was still network television after all—and subject to all of the usual FCC guidelines.

Moreover, from week one forward, sports reporters through-out the nation denigrated the league, calling it "fake football" and worse. Many implied that the involvement of WWF's Vince McMahon assured that the games were fixed and not to be taken seriously—just like the then current incarnation of pro-wrestling.

By March 31, the NBC telecast of the match between the Chi-cago Enforcers and the NY/NJ Hitmen, the home teams of the league's two most populous markets, set a new ratings record. The event received a 1.5 rating—the lowest audience share for a primetime network television broadcast ever.

Shortly thereafter negotiations began as NBC frantically worked to get out of their multi-year commitment to the already floundering sports franchise, and on May 10, 2001, three months after its audience-grabbing (and alienating) debut, the network announced that there would be no second season for the XFL.

Though several of the players did go on to the NFL, none ever achieved any level of stardom. Perhaps the only lasting contribution the project made to broadcast sports (besides providing the perfect example of a bad idea and the nadir of sports broadcasting) was the use of the so-called sky-cam to provide broadcast with an up-close overhead view of game play.

Malpractice Assurance

Few things are more important than health and few fields have seen stranger ideas than Medicine. In the last few centuries we have been able to combine the desperation of trying anything that might cure you with the absurdities of modern technology.

High-voltage Medicine

E. J. Neiburger

In 1888, Nikola Tesla invented a new form of electricity called alternating current. He also invented a high-voltage generator humbly named the Tesla Coil. This electric device was made from a series of two nested coils of insulated copper wire with an electric vibrator switch that rapidly opened and closed the electric circuit, thus creating high-voltage energy (25,000 volts/ cycles) at low amperage. When this was connected to a vacuum (Geissler) tube containing a small amount of argon gas, ultraviolet light and electric discharges were produced. This created heat (through diathermy), smelly ozone gas, and caused minor neuromuscular contractions when placed upon a patient's skin. The effect was caused by electric arcing going from the tube to the patient's skin and then into the ground.

The phenomenon was descriptively called "The Violet Ray" after the purple color produced in the vacuum tube. The Violet Ray generators originally were made as a two-part system: the electric coils and the gas-filled electrode tube. After 1920, a single-unit machine was sold with the coils and replaceable electrode combined in the form of a 12-inch torpedo-shaped, bakelite unit. It had a cord that plugged into household power (110 VAC-220 VAC). The machine operated with an adjustable screw and an electric vibrator creating rapid voltage fluctuations. This caused the two coils to create high voltage (20,000 VAC) with low amperage. It was tuned, using a screw knob at the base of the unit, to the optimum vibration (voltage) desired. A low-pitched audible buzz signaled maximum power. The electrode glowed with a violet color and produced ozone as the high voltage electricity and ultraviolet radiation crackled off the electrode.

The electrode was then applied to the area of the body where treatment was desired. As the electrode (tube) approached within 1 cm of the skin surface, electric arcing began, varying in intensity depending on how well the patient was electrically grounded. A snapping sound could be heard as hundreds of small bolts of electricity arced from the electrode to the skin causing a prickly sensation. As the electrode touched the skin surface, the arcing diminished and the subject felt a warm sensation. This heat gradually increased, and required the movement (usually circular) of the electrode in order to avoid burning the tissue by diathermy. In effect, the electrode was rubbed on the skin creating strange lights, sensations, sparks, and pungent smells—great theater.

The usual treatment consisted of rubbing the electrode on the skin or "inner" tissues for five to ten minutes, followed by rest. To avoid over-treatment, the Violet Ray generator would overheat after 15 minutes' use and have to be unplugged for a cooling period. There was no limit to the number of times you could reapply the cooled electrode to your body.

In the late 1800s and early 1900s, life was hard, death often came early, and medicine was quack-ridden and brutally primitive. The average man/woman could expect to live to forty-two years of age if they survived childhood. If you were sick or injured, you had a 50 percent chance of being helped or harmed by a visit to the physician. The impressionable population was receptive to many "quack" devices and treatments because, in part, professional health care was incompetent. At this time, germ theory (disease was caused by germs) was hotly debated among doctors. People were constantly looking for cheap, easy-to-understand cures for the rampant tuberculosis, syphilis, typhus, and other infections that emptied out cities and towns across the nation. Two favorite treatments, electric shock boxes (e.g., the Violet Ray) and radium water (radium salts in water) fulfilled this need, becoming very popular for general preventative and curative treatments of medical illnesses. Both systems used new technology (electricity and atomic radiation) in an age where "new" and "scientific" were considered synonymous with "good." Many people also relied on the old proven feel-good remedy of opium and alcohol tonics such as Ma Grass Celery Compound. The opium killed the pain and the alcohol made you feel good. It was unfortunately an addicting and temporary cure.

Numerous companies sold a variety of Violet Ray Generators in a wide assortment of shapes and sizes with prices starting at five dollars each unit (1910 prices). Physicians and dentists could purchase the more powerful, glitzy, and expensive professional models housed in polished oak boxes with numerous chrome knobs and dials (costing $45). In those days, the average worker earned about a dollar a day, so the average family could usually afford the cheapest of these devices that could be used by their relatives and neighbors too. There were dozens of Violet Ray "electrodes" (Geissler tubes) available, each with its own shape, use, and name. For example, there were electrodes for the left and right sides of the mouth as well as a separate tube for the front teeth, all of which were devoted to treating pyorrhea (gum disease). There were electrodes for numbing the nose (named the cocaine tube), for anal insertion (for hemorrhoids), deep vaginal insertion (for cervical cancer), hair growth stimulation, penis enlargement, rosy complexion, etc. If there was an orifice, there was a tube made to fit in it and a mythical series of treatments that one might follow in order to affect a cure or prevent future illness. It all seemed logical, in theory as well as on paper.

The Violet Ray machines were recommended for treatment and prevention of numerous maladies from asthma to zinc deficiency. Instruction manuals included with the machines recommended regular treatments for diphtheria, hair loss, sore throat, consumption (tuberculosis), rectal fissures, toothache, pyorrhea, flat feet, lagging libido, and lower back pain. There was little research into the efficacy of these treatments but, spurred on by lurid and miraculous anecdotal reports, they maintained their popularity for more than fifty years. This was because of the novel technology used, affordability, placebo effect, and some antibacterial properties of the resultant high voltage, UV light, and ozone. Even a stopped clock is right twice a day.

The Violet Ray could be administered by one's physician, dentist, friend—or oneself. It could be applied with or without undressing—an important consideration in Victorian society. It was inexpensive; it tingled, flashed, buzzed, sparked, glowed, made medicine-like smells (ozone), and relaxed muscles (overstimulation of nerve endings by the high voltage). It was relatively safe; no one was ever reported killed or injured. It was just what a fearful, traumatized, diseased, superstitious, and medically ignorant populace wanted. Sales were brisk for more than fifty years and availability ranged from the corner drug store to the big catalog stores like Sears and Montgomery Ward.

In the 1950s, the Federal Food and Drug Administration sued manufacturers for fraud and terminated manufacture and advertising of most Violet Ray generators, with the exception of its use for dermatological treatments (it did help with psoriasis). The Violet Ray became just another historic, quack device, occasionally used by counterculture therapists or erotic sex devotees (the high voltage was applied to nipple rings, etc.). It looked good on paper and felt great, but swindled thousands out of their hard-earned money. Money they thought they spent for a cure, not a show.

"[Medicine is] a collection of uncertain prescriptions the results of which, taken collectively, are more fatal than useful to mankind."

—Napoleon Bonaparte

Radioactivity Is Good for Your Health

E. J. Neiburger

In 1896, an amazing new power was discovered by Emile Becquerel and was called "radioactivity." People quickly embraced the novel concept, certain this discovery could be adapted to improve their health. It seemed logical, to the mind and on paper, that this new energy, seeming so powerful and wonderful, need only be applied to the body in order to cure the ever-present diseases of the times. More than that, radioactivity probably, if not entirely, could prevent disease. Considering the vast pantheon of terrible and generally untreatable conditions (tuberculosis, cholera, syphilis, diphtheria, cancer, etc.) which awaited everyone in the early 1900s, there was a great need for a simple and cheap "cure." What wondrous times befell the people living at the turn of the century. They were just blessed with the benefits of mass production, cheap over-the-counter narcotics, electricity, and now radiation—or so they believed.

By the early 1900s, it was known that naturally occurring

well water was sometimes found with radium dissolved in it. The radium, a radioactive element, and its radioactive decomposition gas, radon (often called "radium emanation"), came from mineral deposits deep in the earth. This radioactive water flowed to the surface in wells and was often found in natural hot springs—the hot springs then in vogue at resorts, spas, and hospitals. The healing effects of these popular hot springs were well known, even coveted. And the discovery of radiation in some of the health spring waters gave credence to the notion that the radiation was probably one, if not the main, source of the curative effects seen in spa bathers. It was not too great a leap of faith to conclude that since the radiation cured diseases, it could also prevent diseases. This was just like the belief of most everyone in those days that the curative effects of hot spring waters were real and medically proven.

At least on paper, it was logical to conclude that if the radiation-bearing waters of the most expensive hot spring spas were beneficial, then applying radiation directly to the body must also be beneficial. The public bought it hook, line, and sinker. Around 1910, U.S. Surgeon General George Torney, M.D., said that relief from gout, rheumatism, neuralgia, poisoning, diarrhea, etc. could be attained from hot springs treatments. Other experts expounded on the effects of hot springs radioactivity in killing bacteria and other microbes (which it did), stimulating cell activity, throwing off evil waste products and other medical buzz word phrases.

Some even claimed that water without radon dissolved in it was "dead water" and useless as compared with "live water" containing health-promoting radioactivity. The radioactivity cured diseases and stimulated increased health and vigor. Radiation was magic and everyone wanted some. The only hitch was that everyone, especially the poor folk, could not go to a spa or hot springs. Unfortunately, the natural occurring radium/radon that leached into the water decomposed and did not remain in the water long enough to bottle, ship, and sell as an efficacious product. The demand was there, but no means of delivery existed.

American ingenuity bubbled up with brilliant ideas on how to solve this problem and save humankind. Though radioactive water was short lived, radioactive ores and minerals, especially when they were concentrated, held their radioactivity for years—thousands of years. By exposing fresh water to radioactive minerals, some radium/radon could be transferred to the water, rendering it radioactive. The treated water could be immediately consumed. Thus began the radon water generators.

One of the first was the Revigator (1912), which described itself as a "radioactive water crock" and "perpetual health spring in the home." It sold for $29.95. The lucky owner would fill the Revigator crock with tap water, then let it stand overnight absorbing the radioactivity from the radioactive minerals in the ceramic walls of the container. The water was ready to be consumed the following day—for day after day. It was self-generating and a long-lasting investment. An entire flood of these devices became popular, such as the Radonite Jar, Curie Jar, Radium Spa, and Vitalizer Water Jar.

Numerous radiation-containing knockoffs like the Radium Eminator, the Thomas Cone, and the Zimmer Eminator used large teabag-like ceramic balls or cones containing radioactive materials which were dropped into a container of water instead of surrounding it. The concept was well accepted. There was even scientific regulation. The American Medical Association (1916–1939) approved these "radiation eminators" if they produced two microcuries of radon per liter of water per 24-hour day. Some did, others did not, and a few products claimed to be radiation generators but did not have or produce any radiation at all. They were labeled fakes or quack devices.

Radium water was only one product. There were many more that you could apply and/or ingest. In the 1920s, you could get radium-containing tablets, bread, seltzer bottles, bags, soaps, suppositories, chocolates, ear plugs, pads for sore muscles, toothpaste (radioactivity fought tooth decay microbes), contraceptives, digestive supplements, beauty creams, lotions, jockstraps (for sexual virility), cigarette holders (reduce the harm from tobacco smoke), cigarette pack inserts, comforters, jewelry, coasters, pillows, greeting (health) cards, and nose cups (the Radium Nose Cup was worn over the nose and purified the air you breathed). Some radiation pads, like the Degnens Radioactive Solar Pad, unscientifically claimed to be recharged by exposure to sunlight. The Ray-Cura pad was advertised as being filled with radium ore and, if applied to the affected organ, could cure cancer, tuberculosis, epilepsy, and other diseases. This modality continued to be sold up to 1965 when the Gra-Maze Uranium Comforter (La Salle, Illinois) and related products were confiscated and production stopped by the FDA (Federal Drug Administration).

Other, more damaging radiation sources were actively sold. Why buy a radiation generator when you could just buy raw radium? The Radithor was a half-ounce liquid solution containing, guaranteed, two microcuries of radium salts. Some people drank several half-ounce bottles a day! With all this exposure to radiation and pie in the sky claims of health and well-being, there were increasing reports of radiation poisoning and sickness.

These cases multiplied, receiving notoriety in the press as tabloid reports and lawsuits multiplied with them. Industrialist and champion golfer Eben Byers bragged that he drank three bottles of Radithor a day to boost his health and golfing. After many years, he developed radiation-induced osteoradionecrosis, which painfully rotted his face, jaw, and throat before he died in April 1932. The lurid tales and lawsuits of the dying Radium Girls—five factory ladies who painted radium watch dials (and licked their paintbrushes frequently), then slowly succumbed to radiation-induced cancer—gave the alarm that all was not good with radioactivity.

Increased reports of X-ray radiation injuries and sickness in overexposed patients and radiologists, plus the illnesses of radon imbibers, gradually educated and turned most of the public away from radiation-based medicines. The use and testing of atomic and H-bombs persuaded many that radiation was harmful. The flood of B horror monster movies featuring radioactive creatures clinched that impression. Just look at what it did to The Thing or Godzilla. Still radiation devices persisted. The Gra-Maze Uranium Comforter (circa 1960) and the Endless Refrigerator/Freezer Deodorizer (which purified the air in your fridge using radioactive thorium ore, circa 1985) were sold up to present times. Today, the Healing Uranium Health mines of Boulder, Montana, are doing a brisk business. Many new radiation devices can be purchased on-line from the Far East (especially Japan).

Radiation as a health promoter had a good run. It shortened or ended the lives of many thousands of people, gave many more hope and some psychological feelings of well-being. Radiation in the form of scanners, diagnostic and therapeutic X-ray machines, implanted seeds and isotopes is still used in medicine, but is now applied more carefully. In fact, half of the radiation we receive each year is from man-made medical devices. Who can say we are or are not harming ourselves in new, enlightened ways? Radium seemed to be a good idea in its time. What "cures" are killing us now? Why should it be so much different in this new century?

"Orthodox medicine has not found an answer to your complaint. However, luckily for you, I happen to be a quack."

—A cartoon caption by Mischa Richter

"X-ray the Feet; It Sounds Really Neat!"

E. J. Neiburger

The discovery of the X-ray has proven a boon to medicine, but X-rays have also had some unusual applications. One of the bizarre applications of X-ray technology was the shoe store fluoroscope. This device consisted of a box, four and one-half feet high and three feet wide, which contained an X-ray tube, a fluorescent screen, a timer, pointer stick, milliamperage dial, and three viewfinders to look inside the box. The subject (patient?) placed his feet in the machine and set a timer, which would energize the 50 KVP X-ray tube. The radiation passed up through an aluminum filter, the feet, the shoes, and showed an image on the fluoroscopic screen which glowed with a yellowish green color. In the image one could see the outline of the shoes and the bones of the feet. The image was in real time and one could wiggle his toes and see the bones move. Usually the subject (the child), his or her mother, and the salesman looked, at the same time, through the three viewfinders. The salesman (there were very few shoe saleswomen in those days) used the imbedded stick-pointer to identify anatomic areas in the image.

The advertising promoting the fluoroscope stated that you could get a better fit when buying shoes if you could look and see whether the foot bones had enough room inside the shoe. A better fit meant more comfort, increased health, and a longer-lasting shoe (reduced wear and tear). The machine was very popular in the 1940s and 1950s. It appeared in over 10,000 U.S. shoe stores, 3,000 British shoe stores, and 1,000 Canadian stores. Very few machines appeared anywhere else in the world.

The shoe store fluoroscope first appeared in the middle 1920s, where several individuals have taken credit for its development and popularization. Clarence Karrer, a technician for his father's medical and X-ray machine supply business, invented an "X-ray shoe fitter" in 1924. The product's design was later purchased by a fellow employee, who started the Adrian X-Ray Shoe Fitter Corporation.

There was also a Dr. Jacob Lowe, who practiced medicine in Boston. He claimed to have invented the shoe-fitting fluoroscope used for X-raying the feet of injured soldiers during World War I. The machine did a fast exam and did not require soldiers to unlace and remove their mid-calf combat boots (which had a large number of laces). In 1920 he introduced the machine at the National Shoe Retailers Association convention. The patent, applied for in 1919, was finally granted in 1927 and the newly named Foot-O-Scope was produced and marketed by the Adrian Company of Milwaukee and renamed the Adrian line. This firm was later absorbed by the X-Ray Shoe Fitter Corporation, which also introduced the Simplex line of scopes.

In 1924-25 a foot X-ray machine called the Pedoscope appeared in Britain. Made by the uniquely named Pedoscope Company of St. Albans, this unit was similar to the American products but remained, without competition, on the other side of the Atlantic Ocean.

The fluoroscopes were a big hit in most stores. The use of X-rays was novel, the images impressive, and everyone wanted to see their shoes and feet wiggle. Children would go into the shoe stores and spend considerable time watching their feet and those of their friends. Because there were three viewers in each machine, this act became a group event.

There were some problems with the fluoroscopes. They were expensive and took up lots of floor space. They diverted attention from the sale of shoes, since everyone wanted to see the X-ray images of their feet and would line up at the machine. Long lines of lookers (especially on Saturday mornings) would slow sales and obstruct other shoppers. Every new pair of shoes that was tried on required a time-consuming look at the fluoroscope.

With few exceptions, using the X-ray shoe fitter was pointless. The fit of a shoe is primarily based on the soft tissue of the foot, not the bones. The X-ray images did not show soft tissue. Every store had to have one, though, since customers thought it was modern and technologically important. Shoe stores that had fluoroscopes made more sales and profits than those that did not. So whether the fluoroscopes were a significant benefit or not, they were a sales gimmick, and the stores bought them by the thousands.

Another problem with the X-ray apparatus was that they produced and leaked radiation—a *lot* of radiation. The 50 KVP X-ray tube used 3 to 8 milliamps, and in the continuous (20-second) mode was a health hazard. Most scopes produced 7 to

14 REM of radiation to the feet for a 20-second exposure. REM is the older system of measuring ionizing radiation. The American Standards Association, in the 1950s, recommended an exposure of not more than 2 REM for a five-second period. Today, the safety threshold is about one thousandth of this number. In addition to the radiation exposure to the feet, scatter radiation to those standing around the machine was significant, often measured at one-tenth REM at a distance of 10 feet.

This was not too much of a problem for the occasional shoe shopper, who might zap herself and her children a couple times a year. It was a big problem for the shoe salespeople, who were exposed to the ever-present radiation scattering from the fluoroscope, which was in use for most of the day. After a few months of this kind of exposure, one could absorb enough radiation to "glow." Though the unwitting public received significant amounts of radiation, few actual cases of disease were linked to the shoe fluoroscopes. One of these cases, a shoe model, had her feet scoped numerous times. She developed radiation burns and eventually had to have a foot amputated. Another salesperson was diagnosed with radiation dermatitis of the foot, a skin condition caused by radiation burns. Surprisingly, no cases of cancer were linked to the shoe store fluoroscopes.

Eventually, increasing concerns about radiation dangers created a public movement to ban the shoe store fluoroscopes. By 1950, many regulations were imposed by most states as to how and when the fluoroscopes could be operated. The shoe store owners did not protest and, in fact, encouraged the disuse of this sales gimmick gone stale. By 1970, most of the states in the U.S. had banned the use of the machines. In 1981, the last shoe store fluoroscope, found in a department store in Madison, West

Virginia, had its plug pulled and was donated to the Federal Food and Drug Administration. It was a great idea for its time. It looked good on paper and played well to the crowds. But in the end, it was less than helpful, finally disappearing into museums across the land.

"One of the first duties of the physician is to educate the masses not to take medicine."

—Sir William Osler

Thalidomide

Brian M. Thomsen

Morning sickness has always been the bane of pregnancy for most women. Throughout most of history, they simply suffered. Then, in the late fifties, a new wonder drug appeared on the market that finally ended the nausea bedeviling "the expecting masses."

Of course, like most other drugs, certain side effects were later discovered. The most common of these included "fatigue and constipation, an increased risk of deep vein thrombosis, pulmonary edema, atelectasis, aspiration pneumonia and low blood pressure," all of which were fairly treatable by other miracle drugs given the times, and none were likely to cause irreversible damage—at least not enough damage to make it worth taking this popular, best-selling pregnancy aid off the market.

Unfortunately, there was also another unknown side effect. The drug was called thalidomide.

Thalidomide was developed by a German pharmaceutical

research firm in the early fifties as a facilitator for the development of antibiotics, a use that was later set aside as being both impractical and nonfeasible given the research data. The only plus side of the research at that stage was that the drug itself seemed to be completely non-toxic on test subjects, even when administered in unreasonably high doses.

Given the loss of its original purpose, researchers had to find a new use for the drug in order to recoup the high cost of development. After a preliminary test in Switzerland, the drug showed promise as an anti-seizure aid in the treatment of epilepsy, but proved ineffective against seizures during the trial study

What followed was an eventual quasi-trial-and-error series of applications from which it was eventually determined that thalidomide could be marketed as a generic sedative with practical applications for the treatment of morning sickness. It was marketed as 100 percent safe—no test animals had been harmed in any of the trials.

The case was open and shut, and by October of 1957 the drug had made its way to the world market under a variety of names.

The way to approval in the United States, however, proved rocky. Several at the FDA were skeptical of the claims of efficacy of the drug and requested additional information. But the law at the time allowed the drug to be distributed for a limited time prior to approval under the auspices of a category called "experimental or investigational use." Under that law the limited time would be reset each time that the drug was resubmitted for approval. This loophole led to numerous case of the drug being submitted for approval, being denied due to insufficient data, being withdrawn from approval process, only to be immediately resubmitted for approval with the new data that had been amassed/produced during the time it had been under submission before. During the entire process of repeated submissions and denials the drug remained in distribution.

One of the excuses/explanations for the lack of data handed over by the drug company was that the actual application to the FDA was being handled by the sales and marketing side of the business because the drug's effectiveness was discovered after its developmental stage and not as part of its initial developmental regimen—but since it had already been proven to be non-toxic at that time, such a conclusion was obviously transfered to its new uses.

The problem, however, was in the lax method that had actually determined the drug to be safe. No rigorous study appeared to have gone into the original declaration that, in turn, produced the determination that thalidomide was non-toxic. The animal studies alone were insufficient, and no long term or extensive studies on humans had even been set up. Instead of good science, it seemed nothing less than an effort to recoup the losses from a failed research project.

As a result, at least in the United States, the drug was never approved for market. Despite this fact, however, 2.5 million tablets had been given to more than 1,200 American doctors during the "approval process," and nearly 20,000 patients received thalidomide tablets, including several hundred pregnant women. Many of these women (not to mention those outside of the U.S. market) soon discovered that thalidomide was not only ineffective against morning sickness (the use for which it was marketed), but that the drug also increased the risk of birth defects. Women began giving birth to babies with abnormally short limbs, "flippered" appendages, eye and ear defects, and large and small intestine malformations. The drug was also linked to peripheral neuropathy in the mother. As a result the U.S. drug laws were rewritten to eliminate the loopholes through which thalidomide (and probably many other potentially harmful drugs) made their way to the consumer before a proper vetting.

After the drug's elimination as a sedative/cure for morning sickness, the company continued to look for an applicable use for the drug itself. Several were indeed found—and in 2006 the Bush administration FDA granted it accelerated approval for certain treatments, none of which had anything to do with morning sickness or the stimulation of antibiotics.

A Mainstream Wonder Drug

Brian M. Thomsen

"The visions were not blurred or uncertain. They were sharply focused. I felt that I was now seeing plain, whereas ordinary vision gives us an imperfect view; I was seeing the archetypes, the Platonic ideas, that underlie the imperfect images of everyday life."

The above quote was taken from an article on LSD and is partially credited with beginning the "turn-on" generation of people who believed in a better perception of life through chemical enhancement.

It did not appear in *Rolling Stone* or *High Times* or even *Mother Jones* or the *Village Voice*.

Its source was a much more mainstream publication.

It appeared in the issue of *Life* magazine published on May 13, 1957, as a part of an article that was a "first person narrative" of an experience under the influence of LSD by Gordon Wasson, a banker with the firm J.P. Morgan.

By the following year many of the elite of New York society were experimenting with the drug, using it to expand their perceptions. These notables included Time, Inc. co-founder Henry Luce and his wife Clare Booth Luce. (According to *The New Yorker* magazine, Mrs. Luce thought that LSD ought to be kept out of the hands of ordinary people. "We wouldn't want everyone doing too much of a good thing," she said.)

A decade after the *Life* article, Congress made the sale of LSD a felony and possession a misdemeanor, and its control and regulation was handed over to the Bureau of Narcotics and Dangerous Drugs.

Two years later psychedelic drugs as a group were classified as drugs of abuse, with no medical value—which did nothing to diminish their popularity among the counterculturalists who had come to embrace them.

Pandora's box had been opened and it would not be closed.

LSD (the abbreviation for lysergic acid diethylamide) was first synthesized from a fungus that grew on rye grain in 1938, by Albert Hofmann working for the Swiss pharmaceutical company Sandoz. Hofmann was hoping that this new drug could be used to stimulate circulation and respiration, but unfortunately the results of these tests were deemed inconclusive or negative in terms of their effectiveness. Its pulmonary effectiveness was far outweighed by its perceptual one.

Five years later Hofmann accidentally ingested or came into interactive contact with a bit of leftover LSD. He soon experienced some of the psychedelic effects of the drug including dizziness, visual distortions, and restlessness. Though not strictly a hallucinogenic, LSD does cause a distortion and/or enhancement of stimulus and input as well as time recognition resulting in an extreme sensory experience. It quickly became the subject of several studies as scientists searched for a use for its newly discovered properties.

A 1950s survey article cited the following in terms of the therapeutic uses of the drug: "Types of conditions repeatedly stated to respond favorably to treatment with LSD and other psychedelics include chronic alcoholism, criminal psychopathy, sexual deviations and neuroses, depressive states (exclusive of endogenous depression), phobias, anxiety neuroses, compulsive syndromes, and puberty neuroses." Later studies added, "Psychedelics have been used with autistic children, to make them more responsive and to improve behavior and attitudes; with terminal cancer patients, to ease both the physical pain and the anguish of dying; and with adult schizophrenics, to condense the psychosis temporarily and to help predict its course of development."

And furthermore: "One additional interesting possibility of therapeutic use was based on the activating or 'provocational' effect of LSD. The drug can mobilize and intensify fixated, chronic, and stationary clinical conditions that are characterized by just a few torpid and refractory symptoms, and it was hypothesized that such chemically induced activation might make these so-called oligosymptomatic states [having few or minor symptoms] more amenable to conventional methods of treatment. By far the most important use of LSD was found in its combination with individual and group psychotherapies of different orientations. Its effectiveness is based on a very advantageous combination of various aspects of its action. LSD psychotherapy seems to intensify all the mechanisms operating in drug-free psychotherapies and involves, in addition, some new and powerful mechanisms of psychological change as yet unacknowledged and unexplained by mainstream psychiatry."

Thus not only was the drug a tool for elites to, in the words of Aldous Huxley, "open the doors of perception," but also for combating/treating mental illness.

In addition to all of the above benefits, a 1963 report concluded: "LSD has a wide safety margin and in the hands of experienced investigators does not produce hazardous side-effects."

And there was the rub: ". . . in the hands of experienced investigators."

"Experienced investigators" means trained professionals. In other words, "kids, don't try this at home"—and that includes overgrown kids, or as they like to call themselves, mature adults as well.

Given the fact that the drug directly affected one's perception of outside stimulus, it was important to control that stimulus to assure that the subject would encounter minimal hazards. This was hard enough to do in a laboratory or a hospital, let alone at a Park Avenue party or in a Greenwich Village loft. Moreover the drug was also linked to more nebulous "spiritual" experiences, which became harder to explain, let alone control.

Worse, its use was linked to various incidences of unstable behavior, with some specific high-profile cases that led to suicide and accidental death.

Even samples of the drug that had been requisitioned for serious study sometimes found their way into amateur experimentation, so that many who fell far short of the label "experienced investigator" soon found themselves in way over their heads.

The oppressive right-wing backlash against the counterculture movement found LSD an easy target, and quickly labeled it demonic and criminal.

This was a gross overreaction.

It *was* dangerous, however, and its use did need to

be controlled, something which criminalization didn't really allow.

LSD, the wonder drug of the mind, had been labeled a demon drug of the soul, and its original acceptance by the mainstream was quickly forgotten in the wake of the newly inaugurated war on drugs.

That Sinking Feeling

When you build a bad building sometimes it falls down.
When you create a stupid plane it never takes off. Ships, too, have
a way of telling you when something is very wrong. This is called
sinking. As with planes, autos, and weapons, there has always
been a competition to make each new ship bigger, faster, or dead-
lier. To achieve these ends things are occasionally taken
to extremes, often with disastrous results.

"You can't say that civilization don't advance . . . for in every war they kill you a new way."

—Will Rogers

The Courageous Saga of the First Submarines

Douglas Niles and Donald Niles, Sr.

Since the days of the ancient empires of the Mediterranean, control of the sea has consistently been an important goal of any military regime with access to a sea. Nations that have the strongest navies, such as England during the Napoleonic Wars, or the Union during the American Civil War, have been able to blockade enemy ports, disrupt or inhibit enemy troop movements and supply lines, and deposit armies along coastlines with a freedom of movement that states lacking naval prowess cannot employ.

Concurrently, those nations that could not project their power to control the seas have always been seeking ways to impede, neutralize, or cripple the naval might of their more seafaring foes. Ancient tactics included pirates and privateers, as well as blockade-runners and smugglers. Still, what was really needed

was a way for a weaker navy to strike at the powerful ships of
their military rivals.

One idea conceived to solve this problem could not, technically, be termed an idea that failed. The concept was this: put
men in a boat and sink it—but not all the way to the bottom.
Then move the boat through the water, up to an enemy surface
ship, and somehow poke a hole in the hull under the waterline.
Oh yes, also figure out a way to bring the sunken boat back up
to the surface so the sailors can survive.

If anything, the idea looked so absurd to most military leaders that they were convinced it had no real value and that only
a madman, or perhaps a few very drunk sailors, could possibly
be persuaded to even give it a try. Yet, give it a try they did, and
most of them died during the various early attempts. However,
the efforts and sacrifices of the earliest submariners eventually
led to the perfection of one of the most deadly ship types ever
designed.

Although Leonardo da Vinci is credited with the first submarine design during his spectacular career in the 1500s, the
device he designed was never built. Other Europeans, during
subsequent centuries, came up with a variety of means for sub-
mersing and propelling a boat, but none of these was ever put
to practical use. The first useable submarine was eventually
conceived by an inventor serving a weak, desperate country,
engaged in a war with a mighty naval power.

An American, David Bushnell, designed and built a craft he
called the *Turtle* in 1776, at the start of America's Revolution.
He conceived the ship as a means of attaching explosive mines
to the hulls of British warships. With a hull of oak, the Turtle
was topped by a brass-encased conning tower, a propeller, and
ballast tanks for raising and lowering the buoyancy of the new

vessel. Only seven feet high and three feet wide, the sub was operated and manned by a single person. The vessel moved through the water when the operator turned the propeller by one or both hands, while using the rest of his hands to adjust the boat's trim, control its buoyancy, and steer. (An operator with only two hands would, naturally, be kept very busy.)

The ship was intended to approach the hull of a British ship at anchor, allowing the operator to drill a hole in the hull that would then be filled by a keg of powder, which would be exploded by a timed fuse. Unfortunately for Bushnell, the attack proved more daunting than he had imagined. He did approach a warship, HMS *Eagle*, but was unable to remain pressed against the enemy hull long enough to bore his hole, despite trying in several different locations. During the attempt, the *Turtle* sank, but Bushnell was fortunate enough to swim to safety.

The first successful submarine attack occurred during the American Civil War. Naturally, it was the Confederacy, faced with the Union's control of the seas, which made the bold attempt. A consortium of men led by cotton broker Horace Hunley began working in New Orleans, creating a submersible ship named the *Pioneer*, powered by sailors lying prone and turning hand cranks that were something like bicycle pedals. The vessel successfully sank a barge in a test, but had to be scuttled when Union forces occupied that great port city during the early stage of the war.

Hunley moved his operations to Mobile Bay, and began working with a more advanced design. The vessel was called a "David," in reference to the Biblical hero's battle against Goliath. It was powered by some eight men who turned hand cranks to drive a propeller, while a ninth crewman steered the submarine and raised and lowered it in the water. A long pole protruded from the nose of the vessel, and at the tip of the pole was a mine designed to destroy the hull of a target ship. The men worked in very close quarters, with no air supply except that which was captured in the ship when the hatch was closed.

The *David* was transported to Charleston, where it was to be used in attempting to break the Union blockade. In testing in that harbor, the submersible sank not once but twice, the first time with some loss of life, the second with the loss of all the crew. Hunley himself perished in the second attempt. The boat was recovered and renamed for Hunley.

Despite the reluctance of the Confederate commander, General Beauregard, to order the ship into action, some of the intrepid sailors who had survived the first sinking finally prevailed upon him to authorize a third attempt. In February 1864, the ship, now christened the CSS *Hunley*, embarked on a combat mission. Operating very low in the water, it was not truly submerged, but it remained unnoticed as it approached the USS *Housatonic*, a steam frigate at anchor just outside the harbor.

The torpedo exploded and the *Housatonic* went down, becoming the first warship ever sunk by submarine. Unfortunately, the resulting blast—or perhaps the wake of a passing steamer—proved too much for the not-terribly-seaworthy *Hunley*, and the submarine was lost with all hands. (The vessel would be recovered again, but not until 1995, when it would earn well-deserved status as a museum piece.)

There would not be another successful submarine attack for some fifty years, but the world had taken notice. Every major naval power began to develop submarine technology, with improvements in propulsion—including battery and steam propulsion—and weaponry, most notably in the self-propelled torpedo, bringing the submarine into the forefront of sea warfare.

During World War I, German submarines came close to bankrupting Great Britain, despite England's clear superiority in surface ships. During World War II, the same German tactic nearly brought England to her knees, and the strategic pressure of the American submarine blockade was instrumental in the defeat of Japan. By the time of the Cold War, nuclear-powered subs could remain underwater for months at a time, and carried enough ordnance aboard one ship to devastate ten or twenty cities in a matter of minutes.

Indeed, the submarine has come of age.

"The Nation that makes a great distinction between its scholars and its warriors will have its thinking done by cowards and its fighting done by fools."

—Thucydides

The Sinking of the *Vasa*

John Helfers

The ill-fated maiden voyage of the Swedish ship-of-the-line *Vasa* is a case study of how not to build a seventeenth-century warship. Internal problems and external pressures combined to make what should have been the flagship of Sweden's navy end up at the bottom of Stockholm harbor.

In 1620, the Swedish Navy consisted of about one hundred old, small, lightly-armed vessels. Wanting to protect his states (modern-day Sweden, Finland, and Estonia) with their long Baltic coastline, King Gustavus Adolphus saw that his nation's power could be projected by sea. He commissioned four ships, two 108-foot vessels, and two 135-foot-long ships, from the Stockholm shipyards in 1625. The shipyards, run by two brothers, Henrik Hybertsson and Arendt Hybertsson de Groote, were already in economic trouble, which delayed construction. Several months later, the Swedish navy lost ten ships in a storm, and King Gustavus sent a letter to Admiral Klas Fleming telling him to make sure the Hybertssons built the ships as quickly as possible.

Along with the message, Gustavus sent measurements of the ship he desired, which was the first problem, as the king's plans called for a 120-foot ship. Henrik, however, had already cut timber for a 108-foot boat. After a stern letter from the king in 1626 demanding the 120-foot boat, Henrik added a section to the original plans, making the new ship 135 feet long, which still wasn't what the king had requested, but apparently bigger was better, since construction continued afterward without too much royal interference.

The next problem besetting the *Vasa* was the death of the primary shipbuilder, Henrik Hybertsson. At this time, there were no set designs or calculations for any ship; each one was built based solely on the experience and knowledge of the master shipbuilder. When Henrik fell ill toward the end of 1625 and died in 1627, after a year of construction, oversight of the project passed to Hein Jacobsson. During Henrik's illness, authority had already been shared between the two men, with the lack of central authority creating confusion among the workers.

During construction, several factors combined to ensure the ship's fate. Sixteenth-century warships tended to be unstable by their very design, since most were built with high aftercastles, allowing soldiers to fire upon their enemies from above. The gun decks also followed the wale planks, thick boards that made up the sides of the ship, which often curved sharply upward at their ends. Later designs flattened the decks and cut ports in the wale planks, but that was not the case with the *Vasa*.

King Gustavus played his part in the disaster. In addition to his request for a longer boat, which forced a redesign on the fly, he also requested two gun decks. This was done, but instead of placing the heaviest cannons on the bottom levels, and using lighter cannons higher up, heavy cannons were placed on both levels in an effort to make the ship the most powerful in the world, adding even more weight.

On its first voyage, the *Vasa* didn't carry enough ballast to counter her weight above water. However, even if more ballast had been added, the ship would have sat so low in the water that she would have flooded when the lower gun ports were opened anyway.

Once the boat was finished, Admiral Fleming tested the *Vasa*'s stability by the simple method of having thirty men run from one side of the boat to the other, which would help gauge the boat's tendency to rock. After only three runs, the ship leaned over dramatically, and Admiral Fleming stopped the test, fearing that the boat would capsize. Despite this alarming occurrence, preparations continued for the ship to be outfitted and put to sea.

Finally, when the captain of the *Vasa*, Söfring Hansson, ordered the ship to sail for the naval station at Älvsnabben, the ship set out with open gun ports, which was not the practice of the day. Especially on maiden voyages, all ships typically sailed with closed gun ports so that the captain and crew could get a feel for how she handled. For some mysterious reason, this wasn't done for the *Vasa's* first—and only—voyage.

On August 10, 1628, with a calm sea and a light wind from the southwest, the *Vasa* was towed to the south side of the harbor, where three sails were set, and the ship headed east. After traveling less than one thousand meters, a wind gust tipped the ship onto her port side, where the open gun ports immediately took on water. The once-majestic ship-of-the-line quickly sank in about thirty-two meters of water roughly one hundred twenty meters from shore, killing approximately thirty sailors.

When he heard the news, the king angrily demanded that whoever was responsible be punished. An inquest held at the time found no one guilty—Captain Hansson swore that the guns were secure and that the crew had not been drunk at the time of departure; the original designer, Henrik Hybertsson, was dead; his brother, Arendt, had left the country for Holland; and as for the design changes the king had requested—well, who was going to punish the king for his faulty plans?

The *Vasa* fiasco cost the Swedish government more than forty thousand *dalers*, or about two percent of the nation's gross national product for the year. That comes out to about one thousand *dalers* per meter traveled—a very high price to pay (not to mention the lives of the thirty unfortunate sailors) for such a short trip.

Don't Blame It on Steam

Douglas Niles and Donald Niles, Sr.

A New Look at the Sudden Demise of the Clipper Ship

The clipper ships of the nineteenth century represented the pinnacle of the Age of Sail. They were beautiful and slender, impossibly tall, bristling with masts and bow- and stern-poles in order to put as much area of sail up as was physically possible. They routinely sailed twice as fast as a typical sailing ship, and were treasured as a means of hauling spices from the Orient to Europe, as well as passengers across the oceans of the world. For a short period they attained legendary status as they carried adventurous Americans from the East Coast around the stormy tip of South America to the burgeoning gold fields of California during the 1850s.

By an accident of natural resources, the clippers were uniquely American in design and construction, though many English and a few other European shipping firms purchased and used them as well. They required the tallest possible masts in order to hoist their massive sail surfaces, and these masts came from the virgin forests of New England. Because the clipper builders wanted to mount a third mast, they needed longer than conventional hulls, and these, too, required lumber from untapped, tall-tree forests.

The name and the design both seem to have been inspired by a Baltimore shipbuilder named Thomas Kemp. He built a ship called the *Chasseur*, which had two masts, each hoisting many sails, and a long bowsprit sail extending from the prow. Under the command of Captain Thomas Boyle, the *Chasseur* sailed directly across the Atlantic shortly after the War of 1812 broke out between the United States and Great Britain.

Boyle wielded his fast ship as a privateer, boldly raiding English coastal waters and wreaking havoc with British shipping even as the powerful Royal Navy was shutting down most commerce along the American coast. (Boyle reportedly released a captured merchant captain and required him to post notice on the Lloyd's of London office, claiming that Great Britain was under blockade by the *Chasseur*!) In response to the activities of Boyle and other raiders, the Admiralty recalled about half of its ships to home waters, easing the pressure on United States ports.

As a sidelight of history, the British used the portion of their fleet that remained in American waters to stage a raid into Chesapeake Bay, one of the objectives being the destruction of Thomas Kemp's Baltimore shipyard. Although the redcoats managed to occupy Washington, burning the White House and both houses of Congress, they were stopped at Fort McHenry at the mouth of Baltimore Harbor—the battle memorialized for all Americans by the rocket's red glare in Francis Scott Key's "Star-Spangled Banner."

Paintings of the clipper ships that followed the war show a number of elements coming together to create a vessel of exceptional beauty, impressive size, and sleek, streamlined lines. The hull was long and narrow, with a sharply raked prow to cut through the water. In addition to the tall masts, usually numbering three, additional booms and sprits bristled from the hull, each draped with additional canvas.

Because of the narrow hull, clippers were built for speed, not for carrying large volumes of cargo. They were inefficient at hauling heavy freight, and while a lightly armed clipper could make a good raider—as demonstrated by the *Chasseur*—they were no match for a broadside battle with a man of war. Still, they were ideal for hauling passengers, and low weight/high value cargoes such as spices, silk, tea, and mail. It was said that the owner of a clipper could pay for the cost of the vessel with the profit made from two trips to the Orient for spices.

For some forty years, from 1815 to 1855, clippers were the queens of the sea. Their decline, about the time of the American Civil War, is generally attributed to the rise of the steamship. Even though clippers were frequently faster than steamships, they could not be scheduled as reliably—since they could be becalmed, like any sailing ship—and their popularity gradually began to wane. Contrary to popular belief, however, it was not the steamship that did in the clipper but, specifically, two events occurring in 1869 that almost immediately relegated the clipper to historic, rather than practical, status.

The first of these was the completion of the Suez Canal. This new route dramatically shortened the distance from England to her colonies in the east by effectively removing the circumnavigation of Africa from the distance of the trip. However, the canal could only be used by steamships—sailing vessels, which needed to tack into a headwind, could not navigate back and forth in the narrow canal. Thus, the economic advantage and convenience of steamship travel rapidly became apparent between Europe and Asia.

At about the same time, the "Golden Spike" was driven, completing the transcontinental railroad in the United States. For the first time, cargo and passengers could be transported across the nation with relative ease—and shipped more directly over the Pacific—at a price and time frame guaranteed to make a sailing voyage around Cape Horn, even on a fast clipper, a relic of the past.

"A ship may belong to her captain, but the lifeboats belong to the crew."

—Claude Akins, *The Sea Chase* (1955)

The Sinking of the *Titanic*

John Helfers

In an age of seemingly boundless technological advancement, luxury, and privilege, one name, more than any other, symbolized what Mark Twain called "the Gilded Age"—the luxury ocean liner *Titanic*. And it was therefore fitting that this very symbol of an era would also serve to usher in its end, when the ship that was supposed to be one of mankind's proudest engineering achievements sank on April 15, 1912, during its maiden voyage, carrying more than fifteen hundred people to a watery grave.

Although there is plenty of evidence of the chain of human errors that contributed to the sinking—the unsafe speed through a large field of ice, the failure to heed several ice warnings that came in on the wireless, the missing binoculars in the crow's nest—there were also several engineering flaws that, when combined with the aforementioned lapses in judgment, could only result in one outcome.

From its very creation, it seemed that fate conspired against the *Titanic*. The steel for its hull was made in open-hearth, acid-lined furnaces, which didn't remove phosphorus or sulfur, both impurities that weaken cast metal. If manganese had been added to the metal during production, manganese-sulfide would have been formed, arresting the sulfur's weakening effects. Testing of recovered hull samples reveal that it contained twice the safe levels of sulfur and four times the safe level of phosphorus, but only half the recommended amount of manganese. Consequently, when the already brittle metal was subjected to icy cold water—like the North Atlantic that April night, which was a metal- and bone-chilling minus two degrees Celsius—and then sideswiped an iceberg, the cold, weak steel could not absorb the energy of the collision, but fractured under the strain, causing ruptures along more than two hundred fifty feet of hull. Also, the iron rivets holding the hull plates together were substandard, containing slag, or impurities from the creation process. When the collision occurred, these weakened rivets burst along the impact line, increasing the already severe damage.

On the inside, although the ship had been equipped with the latest in electrically powered watertight doors, the bulkheads themselves only extended as far up as E deck. The *Titanic* had been built to withstand four compartments being breached, but when six were opened, the volume of incoming water dragged the bow down, and, like water flowing from one compartment in an ice cube tray to the next, the flooding continued unchecked into each progressive deck, sealing the ship's fate.

When the captain realized the ship was going down, he ordered the lifeboats—sixteen wooden-hulled and four collapsible—launched. Originally, Alexander Carlisle, a managing director at the *Titanic*'s builders, Harland & Wolff, had proposed a design with a new type of davit that could hold a total of forty-eight lifeboats—more than enough seats for the ship's passengers and crew. But the White Star Line's owner, J. Bruce Ismay, opted for a less costly layout, which meant fewer lifeboats, which meant that the *Titanic*'s boats could only hold fifty-two percent of its passengers at capacity. Ironically, England's Board of Trade guidelines stated that all British vessels over ten thousand tons had to carry sixteen lifeboats, plus enough rafts and floats for seventy-five percent of the passengers. Although *Titanic* weighed in at a staggering forty-six thousand tons, the rules had not been adapted for a vessel so large, so even though the ship did not have nearly enough lifeboats, it had more than the number required by the current law.

Even when the evacuation orders were issued, there was no simple way to carry them out. The *Titanic* did not have any kind of alarm system, nor an internal intercom that would have allowed the crew to broadcast messages throughout the ship. Due to their haste in departing, the crew, most of whom had been hired only a few days earlier in Southampton, were not familiar with the ship's layout, and had not performed a single lifeboat drill. In fact, for the first hour many passengers refused to get into the lifeboats, still confident of the massive vessel's impregnability.

Because of this confusion, many boats were launched with much less than their maximum capacity on board—boat number one only carried twelve people when it was launched. If the sixteen wooden lifeboats could have been filled to their maximum capacity of sixty-five people, along with the four collapsible boats that could hold forty-seven people each, hundreds more could have been saved.

The epic disaster brought many changes to the shipping business. All ships were required to carry enough lifeboats for every passenger and crewmember. Wireless radios were also eventually made standard requirement, with twenty-four-hour monitoring. (Although the *Titanic*'s two operators had remained at their posts until the last minute, many smaller ships only had one operator, and couldn't man the radio around the clock.) And the International Ice Patrol was created by Great Britain and the U.S. to monitor icebergs in shipping lanes. But for the *Titanic*'s victims, those changes came too late.

"Perhaps my dynamite plants will put an end to war sooner than your [pacifist] congresses. On the day two army corps can annihilate each other in one second all civilized nations will recoil from war in horror."

—Alfred Nobel

The Sinking of HMS *Hood*

John Helfers

War is the often the catalyst of invention, and the beginning of World War I was no exception. In 1915, the British navy had already been considering a new battleship design to replace their outdated ships, which were often so heavily encumbered with guns and armor that their lower decks were awash in rough weather, allowing seawater to penetrate into gun ports and limiting their use in shallow waters. New requirements for this next generation of combat vessels included a high freeboard, higher-mounted secondary armaments, a minimum speed of 30 knots, and use of the new 15-inch gun system. In short, these ships would be everything the previous generation wasn't.

In April 1916, the Admiralty placed an order for three new battleships, the HMS *Hood*, the HMS *Howe*, and the HMS *Rodney*, each based on a design from E. L. Attwood. The lightly armored, speedy warships would pack a considerable punch with their main 15-inch guns. However, May saw the Battle of Jutland, where three battle cruisers, the HMS *Invincible*, the HMS *Queen Mary*, and the HMS *Indefatigable* were lost to German shells that plunged through their light top armor and ignited ammunition magazines. A fourth ship, the HMS *Lion*, was only saved through fast damage control by flooding the magazines. Acknowledging the threat of this tactic, the new battle cruisers were redesigned, distributing 5,000 more tons of armor across their upper works and reinforcing decks and turrets, and adding additional weapons systems—none of which resolved the problems that would seal the battle cruiser's fate more than twenty years later.

For example, only the forward cordite magazines were moved below the shell rooms, leaving the rest still vulnerable from the top. The newly reinforced deck and side armor still did not protect against shells impacting at all angles. Also, and most important, the deck protection was inadequate almost from the ship's launch. Spread over three decks, the design was intended to allow the outermost deck to absorb the incoming shell blast, with the armor of the next two decks remaining relatively intact. However, the time-delay shells developed at the end of World War I could easily penetrate the upper deck and explode inside the ship. The extra armor provided just enough protection against the heavy guns of the day, but left the ship vulnerable to future advances in firepower. Finally, the additional thousands of tons of armor and weapons meant the *Hood* was terribly overweight, making her a "wet" ship, with lower decks awash in rough seas—one of the problems the Royal Navy had been trying to eliminate in the first place.

Although serious consideration was given to scrapping the already-started design, there was nothing else with which to replace it, so the *Hood* was launched on August 22, 1918, and commissioned in 1920, after the war was over. This also meant that the class of ships the *Hood* was originally built to fight were never constructed—instead, naval engineering progressed beyond the battle cruiser's capabilities throughout the next two decades, while the *Hood* made do as best it could. Because she was practically a flagship for the Royal Navy, she saw constant service, and was therefore prone to much more wear and tear than the rest of the fleet. She received a major overhaul in 1929–30, and another refit had been planned for the end of the following decade, but the outbreak of World War II made that impossible. Except for a minor refit in early 1941, the *Hood* sailed in such poor condition that she was unable to reach her top speed.

When news came that the German battleship *Bismarck* had been launched, the *Hood* and the *Prince of Wales* were sent to intercept it. They found the state-of-the-art German battleship along with its escort, the heavy cruiser *Prinz Eugen*, on May 24, 1941.

From the start the *Hood* fared badly. Approaching at a poor angle, she could only bring her two forward guns to bear on the two enemy ships, while they could unleash full broadsides. The *Hood* powered toward the enemy as fast as possible, to prevent taking deadly plunging fire on her now totally inadequate three-inch top armor. However, the enemy ships found the *Hood*'s range, and began shelling. An eight-inch shell from the *Prinz Eugen* struck the boat deck, igniting armor-piercing ammunition there, causing a fire that threatened the entire vessel. The *Prinz Eugen*, on orders from the *Bismarck*, then turned its guns on the *Prince of Wales*.

Approximately ten minutes into the battle, a gigantic jet of flame erupted from the area of the *Hood*'s mainmast, followed by a terrific explosion that destroyed the aft part of the ship. She sank almost immediately afterward. Only three of the 1,418 men aboard survived.

Although two separate Boards of Inquiry determined that a shell from the *Bismarck* had struck the *Hood*'s aft magazine, detonating it and causing the loss of the ship, alternate theories on the battle cruiser's destruction include that the fire on the boat deck may have penetrated to an internal magazine; extra ammunition stored outside a magazine may have detonated; or one of its own guns may have exploded. The true cause of the sinking of the *Hood*—even after the wreck was located in 2001—may never be known. What can be said is that the ship, even obsolete and in disrepair, fought to the very last. However, her most deadly enemy was the engineering flaws that were built into her, which certainly contributed to her destruction.

"But the power of destiny is something awesome; neither wealth, nor Ares, nor a tower, nor dark-hulled ships might escape it."

—Sophocles, *Antigone*, l. 951

We Shall Never Again Surrender!

Paul A. Thomsen

After the conclusion of the Second World War, France tried numerous plans to regain their imperial prestige. They tried to reestablish colonial dominance in Africa and Vietnam. They periodically sought to undercut the long-term goals of their more industrious political and industrial rivals through diplomatic sleight of hand and demonstrations of realpolitik. They even embraced nuclear weapons.

All failed to give them an edge over their neighbors. Yet, in the aftermath of the Cold War, while other European nations downsized their navies, France committed itself to a plan that ensured regional hegemony by building and launching the "biggest and baddest" aircraft carrier to sail the North Atlantic, the FNS *Charles De Gaulle*.

An innovation of the post–First World War era, aircraft carriers had been used to great effect by various nations to project power far beyond their territorial borders in both war and peace. During the Second World War and throughout the Cold War, the United States and the other great nations utilized these mobile artificial runways on the rolling seas as landing strips and launching platforms for reconnaissance missions, first strike operations, enemy infiltration, search and rescue operations, and humanitarian relief expeditions. Having invested much in the flawed Maginot Line and suffering the embarrassment of the subsequent occupation force of Nazi Germany, France only looked to the sea as an expression of postwar power. After the war France rebuilt its military and maintained a small carrier fleet.

When the Soviet Union crumbled, however, their allies engaged in a program of military de-escalation—but instead of following suit, French pride demanded she at least maintain the status quo. In 1980, the French Defense Council had authorized the creation of two new nuclear-powered carriers, but even after the fall of the USSR they moved ahead with the building of the two ships, though it would be a long time before either could be put to sea.

In April 1989, the keel for the first vessel (French President François Mitterand insisted it be named the *Richelieu*) was laid. In the early 1990s the project faltered under economic pressure, but France's two remaining carriers, the FNS *Foch* and the FNS *Clemenceau*, were also getting a bit long in the tooth. Initially, public sentiment turned against the floating symbol—until one French politician engineered an erudite *fait accompli*. Rather than cut their prized naval project, French Prime Minister Jacques Chirac repositioned the matter as an issue far greater than military supremacy on the high seas. Renaming the vessel the *Charles de Gaulle* after the World War II hero and postwar French leader and selling off the aging *Foch* to Brazil (the *Cle-*

menceau had already been mothballed), he argued that building this grand aircraft carrier was now a matter of French pride. The tide turned back in support of the vessel, the critics receded, and the slow-going project limped towards completion.

Now with France carrier-less, the *De Gaulle* would truly have to be the grandest aircraft carrier to sail the seven seas to meet expectations. Indeed, she would have the best of everything, including American surveillance planes (specifically, the E-2C Hawkeye) and catapults as well as two nuclear reactors. She would stand in the water 261 feet in length, at 38,000 tons, a worthy symbol of her country.

But as the ship approached her launch date, everything seemed to fall apart.

First, sea trials were delayed until January 1999 when French officials realized the deck length was far too short to be able to allow for an emergency landing by the fighters they had bought from the Americans. It was hastily extended 4.4 meters.

Next, the deck's fresh coat of paint contained a corrosive agent that ate through the catapult wires used to bring landing aircraft to a complete stop. The wires were hurriedly replaced and the deck repainted.

Third, when the vessel got underway, radiation monitors revealed substandard shielding around the reactors. Additional plating was added.

Fourth, when the vessel entered the Bermuda Triangle en route to its Caribbean destination, one of the nineteen-ton port propellers snapped off and sank to the bottom. Evidently the vibrations engendered by the carrier's engines were more than the propellers could handle, and when the navy sought out the original manufacturer who had cast them for inquiry on the matter (and the possible accusation of substandard workmanship), they discovered that the company had gone out of business back in 1999. Ergo retooling the previous part was now out of the question, and the possibility of further weakened propellers negated the possibility of using the replacement stock.

After the *De Gaulle* limped back to France, the broken propeller was finally replaced by a cannibalized prop from the mothballed *Clemenceau*. It was determined that the old model would safely withstand the vibrations. Unfortunately, the change also resulted in a curtailment of her maximum speed, allowing her a top speed ten knots per hour *less* than the old *Foch* regularly managed.

Since the completion of sea trials, the FNS *Charles De Gaulle* did finally complete the plan her Cold War originators had laid out. She served with distinction in the Afghanistan conflict, sat off the coast of India during the 2002 crisis with Pakistan, and has been used as a symbol of French displeasure by being ordered to sail out of the Middle East in the runup to the United States' launching of the Iraq War.

French pride had been vindicated!

Yet, given the lackluster early performance of the *Charles De Gaulle*, not only were plans for the second carrier dramatically scaled back, but France also incurred the ridicule of its neighbors. With papers such as *The Telegraph* and *The Irish Times* calling the ship the "French Calamity Carrier" and "jinxed," the nuclear-powered vessel became the laughingstock of the international community.

The cost of one nuclear-powered carrier: Seven billion pounds.

The time required to make the carrier sea-worthy: Fifteen years.

The value of French pride: Priceless.

No Plan Ever Survives Contact with the Enemy

No endeavor brings out the best and worst in men and women than war. It has also often been the same for those who create the tools with which battles are fought. Under the intense pressure of a nation at war, technologies can jump forward. Brilliant minds turn to mayhem with amazing results. Occasionally the needs of war lead to designers and inventors spending valuable and limited resources on things that in more reasoning times they would likely have instantly realized were mistakes.

> "I'm all in favor of keeping dangerous weapons out of the hands of fools. Let's start with typewriters."
>
> —Solomon Short, a cartoon character created by R. Crumb

A Whole New Battlefield

Douglas Niles and Donald Niles, Sr.

How Machine Guns Were First Installed in Fighter Planes, Almost . . .

It is a truism of human history that every great development in the area of transportation has, sooner rather than later, been given a military application. It wasn't long after men started riding horses before they were trying to knock each other off their mounts, resulting in a host of innovations—from bridles, saddles, and much later stirrups, to lances and spears—so that they could knock each other off their horses with greater efficiency. No one knows how long it was after the first rowboat was invented before sailors were bashing each other with their oars, then affixing battering rams to the prow so that they could send other men's boats to the bottom of the sea.

Trains and steamships were first designed to transport people from place to place, but the rails soon became the primary conduit for military operations—the campaigns of the American Civil War, in many cases, were about control of rail lines and stations—and the power of steam was employed to build strong, fast ships capable of ignoring the wind and exerting their owners' power to gain control of the seas.

So it comes as no surprise that flying machines went through a similar evolution. The first men to fly rode up in hot air balloons late in the eighteenth century. Less than a hundred years later tethered balloons were lifting military observers into the sky, where they reported information on enemy troop movements and helped to spot targets for friendly artillery.

When World War I erupted across Europe in 1914, the airplane was still in its infancy, since less than eleven years had elapsed following the Wright Brothers' success at Kitty Hawk. Even so, the lure of powered, controlled flight—coupled with advances in metallurgy, engine design, and general technology—had resulted in a decade of stunning growth in the fledgling industry. By 1908, Orville Wright had demonstrated airplanes to the United States Army, while his brother Wilbur had done the same thing in France.

By the start of the war, many of the major combatants, including France, England, Germany, and Italy, had aircraft designers and manufacturers among their populations, and naturally the efforts of those pioneers turned toward employing their machines in pursuit of national victory.

During the first year of the war, the airplane was used by both sides pretty much as a mobile observation platform. Pilots, often accompanied by observers, would fly over enemy positions, study the ground and the troops below them, and return to base with information that could be used by the generals as they worked toward the carnage that would eventually consume a generation of European manhood. For a time, these fliers remained aloof from the horrors of the ground war; there are many reports of German and French pilots cheerfully waving to each other as they flew off on their missions.

Still, men being men, it wasn't long before a belligerent pilot or observer carried along a side arm, and started taking pot shots at the fliers on the other side. These little plunks were not terribly dangerous, though they undoubtedly had a dampening effect on the camaraderie among the aviators. As the war moved into its second year, airplanes got larger and more powerful, and those ingenious fliers came up with the idea of mounting .30 caliber machine guns on swivels, so that the observer (not the pilot) could take aim at other airplanes that were off to the side or behind the firing craft.

But this was an unsatisfactory (i.e., not very lethal) method of aerial warfare. Given the fact that both the shooting platform and the target were moving fast, and through three dimensions, most likely in different directions and at different altitudes, it was really, really difficult for the gunner to hit anything. What was needed, many pilots realized, was a gun that was lined up with the airplane's course, so that, by aiming the airplane, the pilot could aim the gun at the same time. Such an orientation would, in effect, cut in half the movement variables between the firing platform and the target.

The seemingly insurmountable problem with this idea, however, was the airplane's own propeller. It was spinning away right in front of the pilot's face, a clear obstacle to a stream of machine gun bullets delivered out the front of the airplane. Not only was there no design that looked good on paper, but in fact the whole idea looked so bad on paper that no one was even

willing to give it a try. (Of course, some airplanes had two engines, one mounted on each wing, but these planes were by
definition too heavy, lumbering, and unmaneuverable to recommend themselves as fighter aircraft.)

A young, highly experienced French pilot named Roland
Garros came up with an idea. Garros flew an advanced airplane
made by Morane-Saulnier. It was a trim, fast monoplane (single
winged aircraft) with the wing flush with the top of the fuselage. It was powered by a rotary engine, which is an engine that
has the crankshaft anchored to the frame of the plane while
the pistons themselves and basically the whole rest of the engine rotate in a dizzying fashion around the shaft, and spin the
propeller with them. There was just enough room between the
cockpit and the prop for a small machine gun to be installed. Of
course, the stream of bullets would shoot through the circle of
the rotating prop, and sooner or later (almost certainly sooner)
the slugs would chew away the propeller, causing the plane to
crash and the pilot to die.

But Garros was inventive, brash, brave, and intelligent—the
virtual archetype of a fighter pilot—and he had an innovative
idea. Enlisting help from engineers on the ground, he put his
idea into effect: two steel plates were bolted to the propeller at
the location where the bullets would strike it. The plates were
canted at an angle to deflect the bullets away from the aircraft
and the engine, and were carefully balanced to keep the propeller true.

After trying out his idea on the ground—it worked just as
he thought it would—Garros took his plane into the skies. In a
matter of weeks he had shot down five German airplanes, and
became a true war hero in the eyes of the French public. The
newspapers dubbed him an "Ace of a man," introducing a term

that is still used to describe a fighter pilot who has shot down five enemy planes in battle.

Unfortunately for Garros, even a modern World War I airplane was still a fairly primitive device. It was not his invention but simple engine failure that proved his undoing. Lacking power, he glided to a crash landing behind enemy lines. Realizing the importance of his invention, he tried to burn the aircraft, but the plane vexingly refused to catch fire.

The Germans recognized the plane, and quickly called in famed Dutch airplane manufacturer Anthony Fokker to study the feature and see if he could duplicate it. Though there is some controversy regarding exactly who thought of this idea or that idea, and who did what first, the practical outcome was that Fokker came up with a better system: He synchronized the position of the propeller to the firing pin on the machine gun, so that the gun wouldn't shoot when the prop was right in front of the barrel. He mounted his new device on another monoplane, one of his own design called the Fokker Eindecker—which was much like Garros's Morane Saulnier. The plane went on to dominate aerial combat for the next ten months, a period of the air war that became known as the Fokker Scourge.

Garros, meanwhile, escaped from a German POW camp and returned to action. By that time, the Allied powers had discovered Fokker's secret and incorporated it into their own aircraft. Garros returned to action, flying a Spad with two forward-firing machine guns synchronized to avoid the propeller. He was killed in action two months before the war ended in November 1918. The idea of synchronized machine guns continued through several generations of aircraft development, and several more wars, until advances in jet propulsion finally rendered it unnecessary.

> "Any community's arm of force—military, police, security—needs people in it who can do necessary evil, and yet not be made evil by it. To do only the necessary and no more. To constantly question the assumptions, to stop the slide into atrocity."
>
> —Lois McMaster Bujold, *Barrayar* (1991)

A Battle Strategy That Will Take Your Breath Away

Brian M. Thomsen

World War I had become quite boring.

The Germans and the Allies found themselves in a face-to-face showdown across France. Unable to make progress through the deadly hail of enemy fire, each side continuously bombarded the other from their positions in fortified trenches and embedded artillery positions, quickly turning everything between into a barren wasteland. Despite the ferocity of the attacks, the battle became a drawn-out stalemate.

It soon became clear that a new factor had to be introduced to disrupt this status quo—but what could it be? They needed "a weapon so appalling that it would destroy not only an enemy frontline but also the will to maintain troops on that frontline."

Such a weapon would force the enemy to abandon their secured positions—perhaps even cause them to turn tail and run—and revolt against the high command that tried to keep them in place. Once they had left, their positions could be supplanted and eventually victory would be achieved by breaking the stalemate.

The enemy needn't be killed, just incapacitated—and if this could be achieved without even having to get up close, so much the better.

But what weapon would provide this solution?

An answer was devised—poison gas.

There is still a considerable amount of debate as to who used "poison gas" tactically first. What is not in dispute is the effect that the gas had on those who came in contact with it.

At a minimum, those targeted had only seconds to don their gasmasks and cover-ups in hopes that they could sufficiently protect themselves during the onslaught. Such protections were quite cumbersome, restricting both mobility and vision and making the actual waging of war quite difficult. Even those thus protected from the gas were largely knocked out of the fight.

And it was critical to gain protection from the effects of the gas. The poison-gas solution provided both sides of the conflict with numerous painful or deadly options.

The most common poison, mustard gas, was primarily a blistering agent. While its effects might not be immediately felt by the inflicted, the resulting damage could be far reaching. Exposure often caused surface blistering on exposed skin, severe eye irritation that could result in eventual blindness, and worst of all, lung scarring from inhalation. Exposure of over fifty percent of one's body was considered fatal, and inhalation could also lead to pulmonary edema. Worse, men who were not involved in the actual attack could still be affected. The insidious gas clung to the primary victims, then often spread to anyone who tried to render aid. Most of these men did not even realize they had been exposed until the later onset of symptoms.

Despite the horrors of mustard gas, it was often survivable. Other gases, such as chlorine-based phosgene, were decidedly more deadly. When inhaled, these gases caused permanent lethal scarring to the lungs usually resulting in an expedient yet painful death. Phosgene's distinctive odor and the gray-green colored cloud it formed upon dispersal made it easy to detect, spreading dread as the entrenched victims became aware of its approach.

But while phosgene was more deadly than mustard gas, it was also more easily neutralized. Instead of a cumbersome suit, sometimes a damp kerchief was sufficient to prevent the gas from entering one's nose and throat. (It was rumored that human urine worked better than water as a dampening agent, as many believed its ammonia content would chemically cancel out the chlorine content—this however proved to be mostly an old soldier's tale as its efficacy is scientifically doubtful at best.)

The gases were usually released by aerosol or bomb/grenade of some sort where the "gasser" would initially dispense the gas and let the wind currents do the rest, carrying it into enemy territory and into the fortified trenches.

Many of the British high command were appalled at the prospect of such a weapon. Commander Ferguson of the British Expeditionary Forces said: "It is a cowardly form of warfare which does not commend itself to me or other English soldiers. We cannot win this war unless we kill or incapacitate more of our enemies than they do of us, and if this can only be done by our copying the enemy in his choice of weapons, we must not refuse to do so."

Eventually, though, Ferguson's contemporaries embraced this weapon as well. But even the careful British, who were slightly late in the gas strategy game in World War I, lacked the necessary control and foresight to avoid the complications of their new weapon. They forgot the old adage of the boy who spit into the wind.

Consider what occurred during the Battle of Loos. General Douglas Haig and the British First Army engaged the Germans in France at what was referred to as the Second Battle of Artois on September 25 through 28 in 1915. Previous to this engagement the British had experienced a certain bit of success at clearing the lines before them through the release of chlorine gas. They would release the gas, which drifted forward over the German trenches and foxholes, thus clearing the way for the Anglo-French forces to proceed onward.

According to the records at hand on the first day of battle, the British followed the German tactic and used chlorine gas to launch their attack starting at around 5:50 a.m. Unfortunately the wind had not been informed of the battle plan and failed to propel the noxious cloud forward. Moreover, on the left side of the British ranks there appeared to be a change in the weather— or more precisely the wind direction. Indeed the breeze happened to change course, turning to propel the gas cloud back from whence it came, towards the British, who were ill prepared for the onslaught.

The remaining force continued forward, oblivious to this development. The main attack began at 6:30 a.m.

Weighed down by heavy packs, the British troops tired quickly against their unseen enemy. The field was soon filled with discarded gear as the soldiers struggled to advance against the gas. Reserves failed to arrive. As a result the British that day lost one-sixth of their forces. Indeed, more Brits died in battle on September 25 than in the entire three years of the Boer War.

Even a fearsome weapon such as poison gas only works when it can be controlled—and then only when Mother Nature is willing to cooperate.

> "Wars may be fought with weapons, but they are won by men."
>
> —General George Patton, Jr.

Tanks a Lot

Bill Fawcett

Among the many areas where excess meant disaster, the armored vehicle, the tank, has been the one most open to abuse. Some designs can be explained by a sharp learning curve, some only looked good on paper.

The Char d'Assault St. Chamond

Bigger is often better in war machines. But a case of tank envy (the Brits had them) by the French Army in 1917 led to the creation of the massive Char d'Assault St. Chamond. Large tanks with crews of a dozen men were doing well in the no-man's land between the trenches and often succeeded in breaking through the German trench and machine gun lines. So the French devised an even bigger tank, almost 27 feet long and close to 9 feet wide, mounting numerous machine guns and one of their superb 75mm guns. The real problem lay not in the armament or the huge size, but in the fact that the treads moving this behe-

moth were only about 16 feet long, leaving the front and rear of
the tank hanging out far beyond them. This meant that the St.
Chamond hung up on any deep shell hole or trench. Once they
hit the front lines these large and expensive armored vehicles
were quickly turned into slow but well armored ammunition
carriers.

Sturmpanzerwagen A7V

The French were not alone in the bigger and dumber tank competition. Even the German Army succumbed to a design that
seemed good on paper but was tragically flawed. The Allies had
used tanks with great success and the German engineers felt the
need to do them one better. The purpose of the tank in World
War I was to break through the barbed wire and silence the
machine guns so that your infantry could advance. This meant
they had to travel over very rough ground and even cross deep
trenches. This fact seems to have been lost on the designers of
the A7V. The 26 by 10 by 11-foot tank held six machine guns
and a 57mm gun that was quite sufficient for blasting apart
those annoying machine gun nests. What this vehicle also had
was a ground clearance of 1.5 inches. Yep, that is not a typo,
the distance between the bottom of the tank and the ground
was a tiny inch and a half. This meant that the entire bottom
of the tank constantly ground itself against the dirt and debris
below it except when running on the smoothest and firmest flat
surfaces. (Maybe the German tank testing area was paved.) That
scraping in turn meant the A7V was slowed to a crawl and was
often stopped by even shallow depressions. There being a lack
of fully paved battlefields in World War I, the Germans found
that their new and expensive tank was effectively incapable of

actually reaching the enemy and could not keep up with even walking infantry on an unpaved road. The program was dropped after only twenty of these awkward giants were made.

T-35, M11, and T-100

In the period between the wars an arms race for tanks began. Rather than bigger, it was decided what was needed were tanks with guns and more guns. The logic must have been that if one large caliber gun was effective, two, three, or even five would be better. The result was the creation of several multi-turreted tanks. The problems of weight and actually manning the guns soon doomed all these efforts, but several actually made it into production.

The Russians had two multi-gun monstrosities made in 1933 and 1939 and the Italians one in 1939. The first of these was the T35 Heavy Tank, and we do mean heavy, which featured three turrets: one raised and two below facing forward and back. It was quickly found that the top turret had to be armed only with a short barreled (and so less effective) gun to keep its fire from damaging the two lower turrets. Also the weight of the turrets meant that the armor on the T-35 could be no thicker than 30mm anywhere and 10mm in most places. At 10mm a good machinegun could punch through it. Eventually the few that were sent to active armor units proved so useless than no more were made.

But the idea of more guns was hard to resist and in 1939 the Russian T-100 and Italian M11-39 tanks were created. This T-100 was heavier, but at least somewhat better armored than the T-35. It also had two guns, with a second gun set below the turret in the hull. Just for good measure four machine guns were also added. The tank was too heavy for most terrain and needed a crew more than twice that of any other Russian tank. A few were actually sent to fight in Finland, where they proved useless. The few that were made did see combat in the 1941 defense of Moscow. Dug in like metal pillboxes, they might have actually been of some use.

Being unable to put enough armor on a tank that had multiple large and heavy guns or a very heavy gun was a problem for anyone who designed tanks. It became an even worse problem when the Italians tried to do this with a lighter, medium tank. The result was the M11/39. This tank only had one gun, but to allow the smaller tank to carry a heavier gun this weapon was built, like the T-100, into the hull of the tank and so had only a limited traverse. This meant it could only fire at targets that the tank was facing. The turret then carried two 30mm machine guns. In order to fit the main gun lower required that the entire hull be heightened, giving the tank a high, flat appearance (easy to hit and easier to penetrate). Further weight was conserved by the M11/39 having a maximum of only 30mm of armor. These tanks, whose gun could only fire forward and which were tall and easy targets, were eventually sent to fight in North Africa. The British destroyed them in droves and by the time the Afrika Korps and Rommel arrived to bail the Italians out, there were none left.

The Sig 33 and Sherman DD

In war you sometimes have to improvise. Two armored vehicles that actually saw combat were very much a case of trying to solve a problem with what seemed to be a good idea but wasn't. The Germans' style of mobile warfare created a need in 1940 for artillery that could keep up with their panzers. Not only was artillery slow to move, but it had to be set up in each location for use. So it must have seemed a stroke of genius to literally put each gun in its own mobile platform and fire from it. This combination of armored platform and artillery piece was the SIG 33. Basically the armored vehicle was a tank chassis onto which was placed a thin metal shield and a 150mm gun. Now the gun was not installed or mounted into the Sig 33, but rather simply fastened on top of the vehicle, wheels and all. The entire hybrid was heavy, slow, vulnerable, and when the gun fired it bounced around on the chassis. The design was quickly dropped when true mobile artillery appeared.

The original need for armor was to assault the fixed defenses created by trench warfare in World War I. By 1944, the need was the same but the location had changed. Tanks were now needed to come ashore with the first wave landing on the defended beaches of Europe to help break through another set of fixed defenses, the Atlantic Wall. But how to get the tanks to the beach? Any boat capable of carrying the heavy armored vehicles was too large to get close to the shore. The solution was basically to put an inner tube around the tank. These Sherman tanks were modified to have a duplex drive (hence the DD designation) that meant they had both normal treads and also a propeller. This meant the tank could move through water toward the beach and be dropped off far enough out that the boats could take them there without hitting bottom. There was of course another problem. Tanks do not float and are not submarines. It was decided that a flotation screen could be put all around the tank, acting basically like an open-topped life vest, and this would keep the tank dry and floating. The problem was that the flotation devices did not work except in the very placid waters of the lakes where they were tested. They were simply not strong or high enough to resist or keep out waves of any size. These same waves or current also made them nearly impossible to maneuver in the water. The Normandy landings were made when the sea was still rough due to a storm the day before. Many DDs did not survive entering the water when being deployed and those that did were quickly swamped. None made it to shore.

The Line Must Be Drawn Here

Jaki Demarest

The Maginot Line, which has become synonymous with the notion of ludicrously ineffective protection, was supposed to render France all but impervious to invasion in World War II. The theory was classic. "If you entrench yourself behind strong fortifications, you compel the enemy to seek a solution elsewhere" (Carl von Clausewitz).

Oh, stop laughing. This really did look good on paper—sort of. Parts of it did, anyway. No, really.

Named after André Maginot, the French Minister of War whose political weight pushed the plan through a somewhat reluctant legislature, the Maginot Line was an extensive and interconnected series of concrete fortifications that France built along its borders between 1930 and 1940. It was built in two sections; the line facing Italy was referred to as the Alpine Line, the line facing Germany was the Maginot, and the name "Maginot Line" was also used to refer to the entire series.

There were initially no fortifications placed along the border with Belgium, as Belgium was assumed to be a staunch and reliable ally with whom the French had signed a treaty in 1920. The Forest of Ardennes in particular was believed to be so dense as to be impassable by a conventional army, a natural fortification of its own with no need of further protection.

This blunder in the initial conception would prove to be the Line's rather rapid undoing.

In 1936, when France failed to challenge Hitler's refortification of the Rhineland (in defiance of the Treaty of Versailles), Belgium decided France was unreliable and abrogated the treaty, declaring its neutrality. France hastily extended the Maginot Line to try to cover as much of the Franco-Belgian border as possible. These fortifications couldn't be built to the standard of the rest of the Line as the water table in the region was high, so underground passages had a tendency to flood.

There was one last massive push to finish and improve the Line in the last two years of its initial construction, 1939 and 1940. It was impressively strong in the industrial regions of Lauter and Metz. If it had been remotely feasible to fortify the entire border to that standard, the Line might well have held. As it was, sections were strong, sections were weaker, sections were wet, and the Line couldn't possibly live up to its own hype.

The Line boasted 108 large forts (*ouvrages* or *gros ouvrages*) at 15-kilometer intervals, each of which housed over 1,000 soldiers, with several hundred smaller forts (*petits ouvrages*) spaced between, housing between 200 and 500 soldiers apiece. Fortresses of the Line were connected by about 100 kilometers of underground tunnels.

Between the forts were interval casemates built to supplement the forts' defenses, manned by crews of 20 to 30. Between the *ouvrages* and casemates, there was a redoubtable barrier of anti-tank and barbed wire obstacle belts. This "line of principal resistance" ran about ten kilometers inside the borders. It was surrounded before and behind by border posts, armored *cloches,* outposts, shelters of interval, quarterings of safety, ammunitions dumps, and observatories, forming a defensive bulwark that was a good 20 to 25 kilometers thick in places. Everything was linked by underground high voltage lines and a telephone network, for steady power and rapid communications.

The *gros ouvrages* in particular were among the engineering marvels of their age. (No, seriously, they really were. This is the part that looked good on paper.) The walls were steel-reinforced concrete, 3.5 meters thick, capable of withstanding tremendous punishment. The larger forts were six stories deep, vast underground mazes, sections of which could theoretically withstand the not-yet-devloped atomic bomb.

Go ahead, reread that, I'll wait. Now think about it and admit it's a thing of weird beauty. Someone drops an *atomic bomb* on you, and you get to come swarming up out of your underground fortresses as soon as the air's cleared a bit, pissed off and loaded for bear. Or, okay, if you're French, you get to smoke a thin black cigarette, curse the triune God, and contemplate the nothingness of being. To each his own.

Propaganda circulated about the Maginot Line was intended to reassure Allied citizens—and it did. Illustrations showed impenetrable multi-story anthills of interwoven tunnels, underground railways, even cinemas. Malls with balls. The propaganda was, naturally, a fair bit better than the reality. Even where the Line was fully constructed, it had weak points that would prove vulnerable to German assault. Ten *petits ouvrages* were captured by the German Army before the Armistice was signed on June 22,

1940, and four more fortresses, including two *gros ouvrages*, were abandoned by their crews, and France's borders were far from completely fortified. But the building of the Line was absolutely consistent with the hard lessons veterans had learned in World War I, in which static entrenchments and defensive combat were the order of the day.

You strategize for the version of warfare you know. Military history is full of examples of strategists, competent or incompetent, who failed to correctly gauge what changing technology and innovation would do to warfare as they knew it. And so it was that World War I veterans, led by André Maginot, one of their own, placed their faith in a static line of fortresses, the strongest they could build. It was enough to convince the French government to spend ten years and around three billion francs making the Maginot Line a reality.

The Maginot Line wasn't without its critics, even in its initial stages. Paul Reynaud and Charles de Gaulle, among a minority of modernists and visionaries, favored investment in armor, aircraft, and mechanized warfare. De Gaulle's tactics were heavily influenced by the lessons of the Polish-Soviet War (1919–1921), in which tanks had made an excellent showing, whereas long lines of static entrenchments had proven disastrous for the Poles. The tide of warfare was changing again, but as usual, those with the shrewdness to interpret the rising trends correctly were in the minority. History, and the dark days ahead for France, would bear them out.

When the Germans finally moved to invade France on May 10, 1940, the necessity of marginalizing the Maginot Line factored heavily into their *Fall Gelb* and *Fall Rot* battle plans. A diversionary force was sent against the Line, while larger forces of ground troops skirted the Line altogether, crossing into France by way of Belgium and the Netherlands, and the supposedly impassable Forest of Ardennes.

Within five days, the Germans had made substantial progress into the French interior, having managed to bypass the Maginot Line almost entirely. The Line, naturally, couldn't move with them. The war hadn't come to the Line, and the Line couldn't go to the war.

A substantial number of French forces had been committed to the Line and were now stuck there, pinned down by the German diversionary forces and entirely cut off from the rest of France in a matter of weeks. By June 22, despite the fact that the Maginot Line was still in the hands of French commanders who wanted to hold out, and the Alpine Line had successfully repulsed the Italians, France signed an Armistice at Compiègne. The French Army was ordered into captivity, forced to abandon their positions within the Line for POW camps.

The unconquerable Maginot Line hadn't been conquered. It had simply been invalidated.

Nor would it prove a particularly strategic playing piece in the Allied invasion of June 1944. The Line, now held by German defenders, was largely bypassed once again; only the fortifications near Metz and in Northern Alsace saw any military action whatsoever.

In the years immediately following the war, the French re-manned the Maginot Line and continued to modify it. The late 1960s saw the Line finally abandoned, as France first withdrew from NATO's military component in 1966, and then developed an independent nuclear deterrent in 1969. The rules of warfare had changed yet again, rendering the Maginot Line more or less completely obsolete. The French government auctioned off sections of the Line to the public, and left the rest to rot (editor).

The End of the Line

> "The bayonet has always been the weapon of the brave and the chief tool of victory."
>
> —Napoleon Bonaparte

A Sword for the Masses

Bill Fawcett

It was not only modern France that has found itself enamored of useless weapon systems. In the era of the French Revolution and the First Empire all men were suddenly equal. In a prior age one of the signs that you were of the better classes was to wear a sword. Perhaps this is why the French Army spent millions of francs to supply the infantry of its army with a short sword called the *sabre-briquet*. The *sabre-briquet* was short, less than three feet long, and simply made with one slightly curved cutting edge and a point. The expected use for this small sword was to give the ordinary French soldier an advantage in hand to hand fighting. And, perhaps arming hundreds of thousands of ordinary citizens with even a small sword appealed to the egalitarian instincts of the ministers.

The reality of this weapon was quite different. To begin, the entire concept of carrying a sword into battle had been made obsolete by the bayonet. Itself more than half the length of the *sabre-briquet*, the bayonet served equally well in hand to hand combat. Further the heavy musket each man carried also served as an effective club and, when the bayonet was attached, as a small pike. There simply was no need for a sword as well. One of the first rules any infantry commander learns is that men discard what they do not need. The Napoleonic French army marched, often very quickly, everywhere it went. Two more pounds of weight in the form of a sword that banged against your hip and had no real use in a fight was just an annoyance. Officers struggled to keep the soldiers from discarding the *sabre-briquet* and often failed. Finally, in 1807, the weapon was abandoned. But even then some of the regiments retained the *sabre-briquet* until 1815. The *sabre-briquet* also looked good when worn by the troops marching in a parade; at least it did to their colonels. And it was useful for chopping firewood if kept sharp.

"In a man to man fight the winner is the one who puts an extra round in his magazine."

—Field Marshal Erwin Rommel

Too Good for Its Own Good

Bill Fawcett

The Germans had their own contributions to absurd ordnance forced on their armed forces. It is normally considered a good thing when a weapon has a high rate of fire. If there is anyone who likes good things in modern weapons it has been the Germans, and the Mauser Company has tried to supply them. So at the end of the nineteenth century, automatic weapons were the cutting edge of military technology and everyone strove to have the fastest firing machine guns and even pistols. But this was the rub, where a machine gun fires rapidly it has a large quantity of ammunition available. Most machine guns are also heavy enough to deal with the constant and strong recoil that firing quickly causes. A pistol has neither of these advantages. But since this was the era of faster firing is better, the Mauser C/96 (1896) "Broomhandle" was created. This is a well-constructed pistol except that it literally fires too fast, expending a clip of six, ten, or twenty rounds almost the moment the trigger is pulled. This meant stopping to reload every time. Another complication was that the fire was so rapid the user was unable to correct his aim while shooting. You could not walk the bullets up to the target, as is the practice with most automatic weapons. Finally the Mauser was light, which made it good to carry, but meant that the rapid fire guaranteed that the barrel would ride up during the burst. Altogether you had one shot, could not correct it, and could not keep a point of aim. This was a wonderful pistol that did everything right—except allow the user to hit anything—and so was near useless in combat.

"The enforcement mechanism for the rules of war is usually more war."

—Solomon Short, a cartoon character created by R. Crumb

Sneaking in the Front Door

Douglas Niles and Donald Niles, Sr.

By 1941, the submarine had long proven itself an efficient, relatively inexpensive, and terribly lethal weapon of war. In addition to the full-size U-boats that the Germans were using to terrorize the convoy routes of the North Atlantic Ocean, small versions of these stealthy attack vessels were employed by several nations for specific, specialized missions. Most famously, the Italian Navy (which had an almost total lack of success employing its powerful surface fleet) sent a midget submarine mission into the heavily defended British Naval base in Alexandria, Egypt. There, these little ships wreaked great havoc by firing torpedoes into British battleships resting at anchor like sitting ducks.

By the summer of 1941, the Japanese had concluded that war with the United States had become inevitable. Imperial Japanese Navy Admiral Isoroku Yamamoto conceived of the strategic strike to commence hostilities: a surprise attack against the United States Navy Pacific fleet, which was likely to be found at anchor in Pearl Harbor, on the island of Oahu, Hawaii. The attack was to be made concurrently with a declaration of war delivered to the American government.

From the beginning, the attack against Pearl Harbor was primarily an air attack. Japan had developed a fleet of fast, modern aircraft carriers, and equipped these ships with some of the most lethal airplanes the world had ever seen. The naval aviators who flew those planes were highly trained combat veterans (of the war with China), and were eager to prove the worth of their new weapons. Yamamoto was more than willing to give them the chance.

At the same time, however, there were officers in the Imperial Japanese Navy (IJN) who were firmly convinced that the submarine would be the decisive naval weapon in the imminent war. With the efforts of the German U-boat campaign as an example, as well as the Italian midget subs in Alexandria, Yamamoto agreed to give this new and revolutionary weapon a chance to strike a lethal blow.

The idea was not exactly embraced by the fliers, who were—presciently, as it turned out—concerned that the submarines might be discovered as they approached the harbor. They feared that the element of surprise could be lost, and the whole attack jeopardized, by a gamble that offered no solid prospect of success. The submariners replied that the midgets stood a good chance of sneaking into the harbor, and that they could launch torpedoes with about twice the explosive power of an airplane-launched torpedo.

In the end, Yamamoto authorized five midget submarines for the operation. Each small craft would be transported to Hawaii underwater, lashed to the deck of a full-sized fleet submarine. The midgets were operated by two-man crews, each armed with two powerful torpedoes. The tiny subs would be released before dawn on December 7, 1941, as near to the mouth of Pearl Harbor as the mother ships could safely approach. They intended to penetrate the harbor, launch their weapons against capital ships—preferably battleships or aircraft carriers—and then sneak back out to sea during the post-attack confusion to a rendezvous point seven miles west of the island of Lanai.

It seems clear that the men who crewed the midget subs did not really expect to survive their mission. They carried pistols and swords, and spoke—and wrote in letters—about landing on Oahu and engaging in glorious last stands of gunfire and hand-to-hand combat. But none of the ten men displayed any reluctance to embark on what was almost sure to be a suicide mission.

The mother submarines all reached positions off Oahu by the appointed time. Between 0100 and 0330 on that fateful Sunday morning, all five of the midgets were released—including one with a broken gyroscope, rendering it almost impossible to control underwater. One by one they approached the harbor mouth, which was anywhere from five to ten miles away from their launching points.

Despite the peacetime conditions and almost scandalous lack of readiness against an air attack, the Pacific Fleet had not entirely neglected submarine defense. A boom with a heavy anti-submarine net protected the narrow mouth of the harbor. The net was some 45 feet deep in a channel that was 72 feet deep at its lowest point. The gap under the net was too small for a standard submarine to pass, though the midgets, with heights of only 20 feet from keep to conning tower, could conceivably squeeze through. A more likely tactic, one that was employed by all of the midget subs accounted for on that bloody day, was to try to follow a ship through the harbor mouth when the boom was opened to allow a surface vessel to pass. In fact, because of morning traffic in and out of the harbor, the net was open for a stretch of more than four hours before and during the air attack.

Additionally, the area just outside the harbor mouth had been declared off limits to submerged submarines—any American sub traveling there would do so on the surface. At least one destroyer constantly patrolled outside of the harbor, and she was authorized to fire upon any submerged sub discovered in that off-limits zone. Additionally, patrol planes, such as twin-engine PBYs, flew over the area and, like the ships, carried live depth charges, and were authorized to use them against any potential targets.

As it turned out, a pair of minesweepers emerged from the harbor at about 0345, and observers aboard one, the *Condor*, spotted something that they suspected was a submarine. The patrolling destroyer, the USS *Ward*, raced to the scene and opened fire with her deck gun and depth charges. The submarine was spotted at or just below the surface, and a direct hit was reported. Even before dawn fully brightened the sky, word of the hostile encounter began, oh so gradually, to filter its way up through the channels of the U.S. Navy Pacific Fleet command.

The *Ward* ended up having a very busy morning, spotting and attacking several other suspicious targets. Observers aboard the supply ship *Antares* also reported submarine sightings, and a second destroyer, the USS *Monaghan*, steamed out of the harbor to join in the hunt. She, too, depth bombed submarine targets. A number of these attacks resulted in oil slicks and even some debris, strongly suggesting that the sailors were attacking more than phantom targets.

Of course, no one made the connection between the underwater infiltration and the potential for a massive air attack
winging down on the island from the aircraft carriers located
nearly 200 miles to the north. By the time the seriousness of the
attacks became clear, bombs and torpedoes were raining down
on the port from the skies.

It has been the conventional viewpoint of history that the
IJN midget submarine attacks at Pearl Harbor were a complete
failure. None of the little craft returned to rendezvous with the
mother ship. Only one of the ten crewmen survived, and his
ship had been sunk without getting off a torpedo; he was captured without a fight, dazed and half-drowned after he washed
up on an Oahu beach. Furthermore, the viewpoint of the naval aviators appeared to have been vindicated—the submarines
before the surprise attack, and only American
lack of imagination served to prevent a gen-
raised throughout the Pacific Fleet.

attempt in Japan to portray the submariners as
a claim that it had been a midget sub, not air
stroyed the battleship *Arizona*. The claim was pa-
served only to irritate the fliers who knew that
they had blown up the great ship. After the war, both American and Japanese historians concluded that the midget subs had
been a misguided and unsuccessful effort.

This conclusion remained essentially unchallenged until
1999. At that time, five U.S. naval officers employed modern
digital photo analysis to examine a photograph of the harbor
taken by a Japanese pilot at the height of the attack. The officers
found conclusive indications that one midget submarine was in
fact within the harbor, and that it fired both of its torpedoes.
They further concluded that one of these torpedoes struck the

USS *West Virginia*, and the other hit the USS *Oklahoma*—two of the eight battleships that were damaged or destroyed during the attack.

Sometimes, an idea which looks good on paper, but seems bad in execution, can turn out not to be such a terrible plan after all.

"God is a comedian playing to an audience too afraid to laugh."

—Voltaire

A Very Low-tech Firebomb Campaign

Douglas Niles and Donald Niles, Sr.

World War II got off to a slam-bang start for the Japanese. With the empire's army bogged down in an unwinnable war in China, the Imperial Japanese Navy took the bull by the horns. Beginning with the attack on Pearl Harbor that destroyed the United States Navy's battleship fleet, the IJN swept to victory after victory—for about six months. The Battle of Midway changed all that: the Japanese aircraft carrier fleet was destroyed, and the inexorable might of the (mightily aroused) American military machine began its relentless advance toward the home islands.

Even before Midway, however, American air power inflicted a grievous propaganda blow against Japan, when General James Doolittle led sixteen B-25 bombers from the flight deck of the aircraft carrier *Hornet*. Flying alone or in pairs over Tokyo and a few other cities, the bombers dropped four five-hundred-pound bombs each—a pittance by comparison to the loads carried by heavy four-engine bombers—on selected targets. Damage and casualties were negligible, and many of the planes and crewmen were lost on the mission. But to the proud Japanese military, the effrontery of this raid was intolerable. As early as 1942, they began to search for ways in which the air war could be returned against the American population.

Meanwhile, in the latter half of 1942 and throughout 1943, the war continued to go badly for Japan. Combining the pressure of naval superiority, the ability to land large armies at positions of their choice, and a steadily growing strength of air power, the Americans closed in on Japan. Submarines cut the island nation off from the resources in its far-flung empire, ambitiously termed the "Greater East Asia Co-Prosperity Sphere." By 1944, bombers based in China were making sporadic, and costly, raids against the home islands.

To combat this wave of modern warfare, the Japanese concocted one of the strangest strategic weapons to appear during the twentieth century. Constructed out of four plies of paper, filled with hydrogen, and carrying something like fifty pounds of explosives, they were balloons. Launched from the northernmost home island, Honshu, the balloons were sent aloft into the newly discovered jet stream, where they were intended to be carried across the Pacific Ocean in about four days. After that time, a fuse was set to ignite the explosives. It was hoped that the resulting blasts would ignite huge forest fires in the vast woodlands of the Pacific Northwest.

Nicknamed *Fugu*, after the deadly Pacific puffer fish, the balloons were launched beginning in autumn of 1944. The first inkling the Americans had of the plan came when the crew of a coast guard patrol boat pulled a mass of gummed paper, attached to something that looked like a bicycle wheel, from the ocean waters off the coast of California. They began to appear with increasing frequency over the winter and spring of 1944–45.

There is no record that a *Fugu* ever actually started a forest fire. The deadliest encounter with the bizarre weapon occurred in May 1945, when an Oregon woman accompanied by five children discovered one of the devices on a hillside during a Sunday afternoon picnic. The resulting explosion killed all six of them, and horrified dozens of onlookers.

The War Department immediately clamped a veil of secrecy over the balloon bombs, successfully preventing any widespread panic. At the same time, the military began posting observers and anti-aircraft guns along the coast. There was some fear that the explosive balloons presaged a more sinister attack, possibly including a biological agent designed to create some kind of plague. By the end of the war, some 17,000 military personnel were engaged in the effort to detect and destroy the approaching balloon bombs.

Eventually, about 9,000 *Fugu* were launched into the jet stream. It is not known how many of them survived the ocean crossing, but estimates suggest that as many as 1,000 made it to North America. American records indicate that only about 30 of them were intercepted and shot down. About 100 were discovered on the ground before the end of the war. Since then, some 150 more have been found, scattered in sites from as far north as the Yukon Territory to as far south as Mexico.

Nobody knows how many of them are still out there, in the deserts, mountains, and forests of the American West. One can only hope that the fuses and explosives are less deadly now than they were during the war, sixty years ago, in which they floated so gently and futilely toward battle.

Nuclear Nonsense

Bill Fawcett

With two nuclear blasts marking the end of World War II, anything atomic had great appeal to the military. Since the Navy and Air Force had bombers and so were nuclear armed, the Army was constantly looking for suitable atomic weapons for their own armory.

The M-65 Atomic Howitzer

This gigantic 280mm gun was nicknamed "Atomic Anne." In 1951, it was capable of firing an atomic shell up to seven miles. What it was not capable of doing was moving. The gigantic weapon required an equally large carriage—in this case one that weighed in at eighty-eight tons. Moving the M-65 was difficult and slow. To move it off a road, you had to bulldoze a new road. Setup took a long time and firing, still longer. Larger even than most railroad guns in World War II, this howitzer was just too huge. If you placed it close enough to hit a serious target, then it was likely to be overrun by the enemy. Eventually the army agreed, and the entire career of the M-65 consisted of one test firing.

Project Pluto

Not to be outdone in creating new and amazing atomic weapons, the Air Force in 1957 began to develop a mach 3 nuclear missile using a fission engine that, in theory, could travel all the way around the world to hit one target. Powered by an atomic engine, the Pluto could stay aloft flying for months carrying a number of hydrogen bombs. The amazing reactor also turned out to be an amazing problem. It was not so bad that the reactor tended to leak. What was worse was that the exhaust of the rocket was filled with radioactive particles. Just flying around guarding a country would do it more harm in the long run than the Pluto's bombs might do to the enemy. Only after significant expense did good sense prevail and Project Pluto was cancelled.

The Davy Crockett

In a way, from its very conception, the Davy Crockett Nuclear Bazooka was a classic military joke. The weapon was designed to fire a nuclear warhead a distance of 400 to 600 meters. The joke came because the blast of the atomic warhead had a lethal radius of about 350 meters, or just over 1100 feet. It was quite possible to fire this weapon correctly and actually still be in the blast zone of the warhead. And yet, this weapon was actually put into production in 1962 and almost 400 were issued to troops. A desperation—read suicide—move, the weapon was only useful if you were going to be overrun, as was the fear in 1962, by hordes of Russian tanks crossing into Europe. Fortunately none were ever used.

The highly secret weapons system did have one real distinction: it was a movie star. While still being classified top secret, some Davy Crocketts were deployed to Okinawa and, somehow, one Crockett, firing a standard warhead, appeared in a Godzilla movie. This was perhaps the only time it was fired in battle—well, sort of a battle. . . .

"Nature abhors a hero. For one thing, he violates the law of conservation of energy. For another, how can it be the survival of the fittest when the fittest keeps putting himself in situations where he is most likely to be creamed?"

—Solomon Short, a cartoon character created by R. Crumb

The Holy Grail of Firearms

Paul A. Thomsen

The M-1 Rifle was a marvel of early twentieth century weapons technology. Developed by Canadian-born John Garand and adopted by the United States Army in 1936, the rifle weighed eleven pounds, held an eight-round clip, could withstand an awful lot of punishment, and, by the end of the Second World War, had become an American infantryman's preferred weapon for dealing death at a distance.

Shortly after the Korean War, the army leadership realized they would need a replacement for the M-1. But the procurement board didn't want just any weapon. They wanted another Holy Grail of weapons technology, and designed review guidelines and test parameters that would ensure their new perfect weapon would be a boon to every future fighting man of the United States military. They believed they had found their perfect weapon in the Springfield M-14 rifle.

As World War II veterans later attested, the M-1 was sturdy, reliable, offered semi-automatic firing, an ejectable "en bloc" clip, and, in a trained set of hands, a highly accurate kill zone to within 460 meters. With other nations fielding small arms with larger clips and the advent of submachine, the M-1 underwent periodic minor modifications (including a fully automatic firing capability) to keep pace with new demands. But, even with the changes, the rifle could not keep up with modern military advances. After Korea, the search for a replacement began in earnest.

During the early Cold War, the army largely took a liking to the M-14, believing it the most likely successor to the M-1. But the model did not please everyone. Improvements had to be made before they would officially authorize the weapon for mass production. What followed was a long process designed to find the perfect fighting weapon.

With the Russians widely fielding the AK-47 Kalashnikov, many in the army advocated their new weapon's "spray and pray" form of fire, but a few more traditional military thinkers favored a continued emphasis on marksmanship. Many remembered the controversy which had erupted in the late 1920s and 1930s over the redesigned ammunition capability of the M-1 from a .276 caliber round to a .30-06 caliber round to meet budgetary restrictions in support of the army's stockpiled World War I ammunition. Budgetary allocations and underutilized stockpiles aside, most studies of World War II combat units showed that the average wartime soldier was reportedly reluctant to open fire on an enemy, but once the firing had begun, the soldier's training and instincts took over, propelling him to retain an aggressive posture. The Pig and Goat Boards (named for their preferred choice of animal test targets) argued the point tirelessly until an outside element, the North Atlantic Treaty Organization's (NATO) weapons protocols essentially made the decision for them.

The M-14 would carry standard NATO ammunition.

Next, they discussed weight and fire capabilities. Ironically, much as they loved the M-1, everyone agreed the rifle had been way too heavy. Hence, any successor to the M-1 needed to weigh no more than seven pounds. Unlike the original M-1 design, their new weapon had to be capable of selective semiautomatic and automatic fire from day one. After nearly a decade and a half of testing, evaluating, changed parameters, retooling, and more testing, the army set its eyes on another Springfield Armory model, the M-14 rifle, as the ideal future small arms weapon. Like its predecessor, it was sturdy and reliable, but it was also lighter than the M-1.

There was, however, one drawback. By demanding a lighter weapon, the army had made room for a new flaw. While debate over the type of ammunition had been settled by agreements with NATO (the ammunition adhered to the new NATO-compliant 7.62 round), soldiers repeatedly struggled to hold the experimental M-14s on their targets when firing at full-auto. It was like trying to restrain a large starving wild animal looking at a herd of plump slow-moving prey; it could, indeed, bring down any one of the herd, the question remained which one and how many passersby it would also kill in the process. That problem was eventually solved by the addition of a pistol grip near the trigger and a second retractable grip further up the barrel, which allowed the soldier to hold the weapon on target more easily.

The debate and retooling of the M-14 over the weapon's finer points continued to rage several years longer, but by 1961, it looked like the army had finally succeeded in crafting the perfect successor to the M-1 . . . that is, until Washington politics got into the mix.

Appointed Secretary of Defense by President John F. Kennedy, Robert McNamara had been tasked with reining in military expenditure. Since taking office, his no-nonsense, curt business style had not won him many friends in the Pentagon and, when his office reevaluated the now fifteen-year-long M-14 development project, many in the army liked him even less. In their pursuit of a new ideal weapon, the military had expended way too much time for the defense secretary's liking. Instead of rubber-stamping the project for rapid production and wide dispersal to American soldiers throughout the world, McNamara effectively killed the project when he chose a less well-tested and developed weapon, the AR-10 by ArmaLite. Originally rejected by the army in the late 1950s, the ArmaLite rifle had been shopped around to Asian marketers (who preferred lighter-framed weapons) and had been picked up by Air Force General Curtis LeMay for his small number of sentries and pilots when the AR-10 caught McNamara's eye. It was light, capable of full-automatic fire, and best of all, it was cheap.

In one fell swoop, political expediency trumped experience in choosing the successor to the M-1. But there would be dire ramifications in the defense secretary's last-minute choice. While new recruits at home were being trained with the AR-10, renamed the M-16, the soldiers using the weapon in combat were already discovering its deadly flaws. By 1962 veteran soldiers touring Vietnam complained that their newly issued M-16s often jammed in combat situations. Inspectors sent to investigate the problem found corroded rifles and heavily pitted barrels caused by exposure to the tropical climate and poor weapons maintenance. Worse, the lubricant bought to maintain the weapon's theater functionality was substandard. Worse still, in looking back over the history of the weapon, it was discovered that the M-16 had performed comparably to its field tests with the M-14, that it exceeded the M-14 in its ability to spray a target, but that it had never been designed to function in a jungle environment.

Oops!

A few angry letters to congressmen, several thousand shipped packages of store-bought lubricant, a number of ArmaLite field kit patches and weapons upgrades later, the army stood by the M-16. While some M-14 variants continued to be utilized in limited instances by special operations personnel several decades later, the perfect weapon so carefully crafted over time never saw the large-scale military deployment for which it had been designed. Instead, because of an act of capriciousness and sheer willpower, the M-16 became the flawed twentieth- and twenty-first-century successor to the Second World War M-1 rifle. Instead of their holy grail, the army got a cracked coffee cup.

> "Before a war military science seems a real science, like astronomy; but after a war it seems more like astrology."

> —Rebecca West

A Heavyweight Too Heavy to Fight

William Terdoslavich

Every new weapons system has to be bigger, faster, better, and more expensive than its predecessor. That is the Pentagon way.

The Crusader XM2001 had many of these attributes. It was an amazing piece of self-propelled artillery—while it lasted. The Crusader's liquid-cooled 155mm gun could fire ten to twelve rounds a minute at ranges up to thirty-one miles. Its computerized fire control system could key the gun to fire eight rounds at different settings so that the shells arrived on target simultaneously. It could sprint 750 meters to a new firing position to avoid enemy return fire.

And it was going to be a real "bargain." You could buy 450 of them for only $11 billion. And each Crusader could do the work of several of its predecessors.

Despite all these goodies, the Crusader XM2001 weighed forty tons. The support vehicle carrying the extra ammunition for reloads weighed thirty tons.

And that seventy-ton package weighed heavily on the mind of Defense Secretary Donald Rumsfeld, the man who lanced the Crusader.

No Fat Artillery Pieces

Necessity conceived the Crusader during the Gulf War of 1990–91. The Army was relying on the M109 Paladin, a 155mm self-propelled artillery piece that started service in 1962 and had seen many upgrades since then. By 1990, the Paladin was outclassed and outranged by more contemporary Soviet artillery, which the Russians sold gladly to Iraqi dictator Saddam Hussein. Fortunately U.S. forces did not have to face an enemy firing better guns while kicking the Iraqi Army out of Kuwait, but the Paladin's performance shortcomings worried defense planners and generals.

The reliable Paladin needed a replacement.

Work on the next gun began in 1994 by United Defense Industries International, a subsidiary of the Carlyle Group. The project was well managed, plugging along on time and on budget throughout the 1990s. Crusader had to be well armored to survive the hazards of the modern battlefield, thus adding to its weight. But two factors conspired to undermine Crusader as it was being turned from a prototype to a weapon. It lacked enthusiastic support in Congress. And the war in Afghanistan, following the September 11, 2001 terrorist attacks, questioned the need for a seventy-ton weapons package that could not be moved easily to distant trouble spots.

The Army was going to kill to get the Crusader.

Donald Rumsfeld wanted to kill the gun just as badly.

The inevitable budget battle was going to be a bloody one.

Lawyers, Guns, and Money

In the spring and summer of 2002, Crusader's proponents and critics fought over the program's fate. Defending the program were then Army Chief of Staff General Eric Shinseki and Army Secretary Thomas White. Shinseki was blunt about the Army's need for the Crusader. Pointing to the war in Afghanistan, the general stressed the need for immediate fire support for troops in the midst of battle. Calling in artillery fire only takes two to three minutes. An air strike can take up to ten times as long, which in combat is like waiting forever.

But the Crusader looked like a fat turkey to Rumsfeld, who complained that 60 to 64 C-17 heavy lift aircraft—half the C-17 fleet—would be needed to move a battalion of 18 Crusaders to the next global hot spot. Each Crusader, fully loaded with fuel and ammo, plus its accompanying support vehicle, would form a transport package weighing 97 tons.

Supporters were quick to point out that the Crusader was never meant to be airlifted. Like much of the Army's mechanized equipment, Crusader would move by ship or would be prepositioned near possible trouble spots to permit more rapid deployment.

However, even heavy equipment can be airlifted in a pinch. NATO commander Gen. Montgomery Meigs made that point. Practicing a rapid reaction force airlift in Europe, he moved 2,500 troops and 325 vehicles to Hungary to simulate a rapid intervention. Sixty-ton M-1 Abrams tanks rolled off of C-5 Galaxy heavy lift transports, along with other pre-positioned M-2 Bradleys and a plethora of HUMVEEs and trucks that were delivered by heavy lift C-17s and the far more commonplace C-130s.

Even Army General Thomas Keane pointed out that a single Crusader could have been airlifted to Afghanistan in a hurry and driven to the front to support U.S. troops in Operation Anaconda, fighting Al-Qaeda and Taliban terrorists near Shahikot.

Rumsfeld then fired back with his heavy gun: Central Command chief General Tommy Franks, who testified before Congress that the single Crusader would not have done much good in Afghanistan anyway.

To counter the Crusader, Rumsfeld stressed the use of Precision Guided Munitions (PGMs) as an alternative. This broad category of weapons encompasses any missile, bomb, or shell that has a guidance system, giving a single round or missile the power to inflict certain death on any target residing at a map coordinate. Why, Rumsfeld argued, use a multi-billion-dollar overweight artillery piece to do the work of a million-dollar smart weapon? His eye was on Excalibur, a light multiple rocket system that could be deployed by a C-130. But Rumsfeld's vision was questioned by Senator James Inhofe (R-Oklahoma). Why fire a $200,000 missile to hit the same target as a $200 artillery shell?

In the end, the weight of Rumsfeld's argument was measured in tons, as he argued that the Crusader wasn't "expeditionary enough." The proof was the deployment of a Marine Expeditionary Unit to secure an airfield in Afghanistan. To speed the deployment, the Marines left their artillery behind, counting on air support to do the same job as the big guns. Franks chose to deploy the Marine unit because it was made to move quickly rather than wait for the Army to move one of its own battalions.

In the end, Rumsfeld got his way.

He killed the Crusader.

And he caused some collateral damage.

Representative J. C. Watts (R-Oklahoma), the number three Republican in the House leadership (and the only African-American Republican in the House), declined to seek re-election because he could not save the Crusader. It was going to be made in Oklahoma.

Shinseki did not improve his standing with Rumsfeld by defending the Crusader. Shortly before the Iraq War, Shinseki testified before Congress that it would take more than double the estimated 150,000 troops to secure the country.

Rumsfeld, who crusaded for the Iraq War, dismissed this argument—and Shinseki, who got replaced by General Peter Schoomaker.

Army Secretary White was forced to resign in disgrace, but that was over his previous involvement with the bankrupt energy-trading firm Enron.

And the Winner Is . . .

The program to develop the Crusader had already consumed $2 billion, with the remaining $9 billion "saved." Crusader's design and performance concepts were rolled into the artillery piece that will become part of Boeing's "Future Combat System," an array of fourteen programs totaling over $160 billion. The concept is to integrate a family of unmanned aerial vehicles mounting sensors with a wireless information network that would provide real-time targeting information to a series of light tanks, APCs (Armored Personnel Carriers), and self-propelled artillery pieces.

The Non-Line of Sight Cannon (NLOS-C) is Crusader's replacement in the FCS program. It will mount a 155 mm gun with a similar computerized firing and auto-loading system as Crusader, but only be able to fire six rounds a minute or four rounds for simultaneous impact. The twenty-ton gun is to be carried into hot spots by the C-130, the most common of the Air Force's transport planes.

The FCS is expected to begin deployment around 2014. Crusader would have gone into service in 2008. But if FCS and the NLOS-C follow the typical path of weapons development, it should arrive late, over budget, unable to work as advertised, and requiring expensive follow-on work to become functional.

That is when the dead Crusader will seem like a bargain.

In 1836, the Creek and Seminole Indian tribes in Georgia and Florida were waging war against the United States. The U.S. Army had its hands full. The 5th Commandant of the Marine Corps offered the services of a regiment of Marines for duty with the Army. Colonel Commandant A. Henderson placed himself in command and, taking virtually the entire available strength of the Corps, left for the extended campaign after tacking a terse message on his office door which read:

Have gone to Florida to fight Indians. Will be back when War is over.

The Double Agent

Paul A. Thomsen

During the Vietnam War, American soldiers faced snipers, saboteurs, formal military forces and a seemingly intractable enemy able to blend into the surrounding Asian countryside. While the United States military bested the North Vietnamese Army (NVA) in every theater battlefield action, the enemy's ability to rapidly appear, strike, and then, just as rapidly, melt away into the thick growth of the surrounding environment consistently placed the western superpower on the defensive.

As the casualty rate for American ground forces patrolling the jungle countryside soared, American war planners and ground commanders looked to a new weapon, called Agent Orange, to help them find and kill the enemy. Once sprayed across designated areas, the herbicide would destroy all nearby plant life and, hence, enable the American military to find the NVA and drive them from the area. It was a sound plan, but, years later, evidence emerged to show that the agent had worked a little too well.

After the Second World War, France struggled to regain control over her colonial holdings in Southeast Asia, but try as they might, the French were repeatedly humiliated by a rag-tag group of Vietnamese rebels. When France finally withdrew, the United States stepped in, determined to prevent Vietnam from being absorbed by Communist interests. The U.S. filled the void at first with a few hundred advisors, and, later, hundreds of thousands of troops. Like the French, they also faced stiff resistance from the unconventional enemy and their ability to use the environment to camouflage their moves. The heat and humidity played havoc with the technologically superior American weapons. The lush swampy terrain populated the countryside with a myriad of ambush possibilities, and what military units could see on the ground fifty yards away, was frequently hidden from air support by thick foliage and dense tree canopies. It was a green nightmare.

In 1962, in an attempt to stem the tide of American body bags and neutralize the growing communist movement, military leaders enlisted the aid of the relatively new agricultural technology of herbicides, hoping to effect greater control over the Vietnamese countryside and expel the NVA. By agreement with the American defense community, civilian chemical companies were contracted to prepare and package different mixtures of herbicides in white, purple, blue, pink, green, and orange color-coded fifty-gallon drums for the military.

Once the herbicides had been transferred into military care,

Vietnamese-based American military units then engaged in an organized large-scale dispersal of the agent throughout the most heavily contested regions in the combat zone. According to estimates, between 1961 and 1975, the military sprayed nearly ten percent of South Vietnam with seventy-two million liters of chemical defoliants by aircraft. An additional six million liters were sprayed in smaller water- and ground-based actions. In little time, the falling clouds of the most widely used herbicide, Agent Orange, ate through much of the green canopy, exposing fields, previously hidden enclaves, and enemy transportation networks to military surveillance and attack units. The frequency of spraying reportedly grew so high that, during Operation Hades and Operation Ranch Hand, one group, the 309th Air Commando Squadron, created a new group motto, "Only We Can Prevent Forests."

Ironically, the military had narrowly defined their enemy as NVA and Viet-Cong (VC) and had, in removing the enemy, inflicted such hardships on the South Vietnamese populace that much of the south also rose up against the United States. Shortly after the 1975 collapse of the South Vietnamese government and the removal of American forces from Southeast Asia, several returning American military personnel were diagnosed with soft-tissue sarcomas or non-Hodgkin's lymphomas.

In 1977, the growing ranks of similarly ill veterans pointed to their frequent exposure to the smog-like mists of Agent Orange as the cause of their illness. Few in government and private business gave serious consideration to their claims. As the years went by, still greater numbers of Vietnam veterans reported incidents of illness consistent with their stricken, dying, and now deceased comrades, all of whom had been exposed to Agent Orange. It seemed that the chemical concoction they had been dumping for years had been eating more than just plant life.

In the late 1970s, media outlet exposés on Agent Orange pressed a few members of the United States Congress into addressing the situation, but the efforts of the few were not able to overcome the intractable position of either the military or the involved businesses, who denied any relationship between the illnesses and Agent Orange. Finally, through the combined efforts of the Veterans Administration (VA), increasing interest by intrepid reporters, and several class action cases, the federal government was forced to reevaluate the effects of Agent Orange. After careful study, several scientists and veterans' lawyers investigating the chemical compounds reached the conclusion that the mixtures were, indeed, linked not only to the defoliated jungles, but also to the ill health of many of the humans exposed to the herbicide.

By 1986, nearly 220,000 Vietnam veterans requested examinations to confirm that their health problems were caused by their exposure to Agent Orange fifteen to twenty years prior. Facing staggering numbers of claimants filing suit and mounting evidence, the civilian companies which had provided the herbicide ultimately settled out of court with the American veterans for tens of millions of dollars provided by the government and the promise of continued medical assistance. Several decades after the American military evacuation of South Vietnam, names continue to be added to the Vietnam Veterans Memorial in Washington, D.C., silent testament to imperial folly and a victims list of America's most insidious double agent.

"If we slide into one of those rare moments of military honesty, we realize that the technical demands of modern warfare are so complex a considerable percentage of our material is bound to malfunction even before it is deployed against a foe. We no longer waste manpower by carrying the flag into battle. Instead we need battalions of electronic engineers to keep the terrible machinery grinding."

—Ernest K. Gann, *The Black Watch*

Sergeant York Misses the Target

William Terdoslavich

It can take years to develop a new weapons system.

It can take months to prove it does not work.

It can take days to kill it.

It can take a minute to roll your eyes in disgust.

The M247 Sergeant York went through all those phases of inept development. The Army needed an anti-aircraft gun for divisional air defense (hence the acronym DIVAD). And they wanted one in a hurry. But the shortcuts taken to meet the need turned into coffin nails that sealed the fate of a good idea badly executed.

Back in the late 1970s, the Army's Vulcan 20mm anti-aircraft gun was getting long in the tooth. The Chaparral anti-aircraft missiles were also aging badly. The Army needed something for low-altitude air defense, especially at a time when the Soviets were fielding new fighter and ground attack aircraft that could easily overcome existing anti-aircraft systems.

In the interest of speed, the Army specified only off-the-shelf components of proven performance would be used for the project. The Army was also going to leave it to the contractors to pull the system together with minimal oversight for a fixed price. They hoped the recipe would yield a useable anti-aircraft gun in only a few years—before rising costs robbed the program of its value.

The result was the Sergeant York self-propelled anti-aircraft gun, which unlike its namesake, could not hit anything. As one press critic put it, the Sergeant York "died of embarrassment." As author Andrew Cockburn added, "It was more of a menace to the taxpayer than enemy aircraft."

The Army began its rush to failure in 1977, when it drafted its specifications for the Division Air Defense system. The replacement had to be made with existing, reliable components. The contract-winning bid put forth by Ford Aerospace and General Dynamics in 1978 took the tank chassis from the M-48 Patton tank and mounted a turret with twin 40mm Bofors anti-aircraft guns. It was then topped by the radar unit from an F-16 fighter and linked to a targeting computer that could keep the guns aimed at low-flying jet aircraft.

Ford and General Dynamics would do the work for a fixed price so long as they had little interference from the Army or the Pentagon. With that arrangement, the DIVAD was supposed to come in quickly and cheaply. The Army wanted to buy 614 DIVADs, deploying 36 of these guns per division. Total program cost would probably come out to around $4.5 billion.

Problems with the program appeared early. By January 1980, the Government Accounting Office (GAO) was flagging the risks found in the DIVAD contract, while Pentagon auditors were taking issue with some of the subcontracting practices and documentation by Ford Aerospace.

Development led to testing.

During testing the DIVAD became a ping-pong ball batted between the Army and Ford Aerospace. Between 1980 and 1983, the Army kept finding problems with the gun that the contractor had to fix. In the meantime, the Army kept begging the Pentagon for one more chance to get the program right. It was an engineering problem that could have been fixed with enough time or money, as the Army kept finding the same problems with the gun.

By January 1983, the Pentagon's Office of Program Analysis and Evaluation rated the DIVAD unready for combat. The Pentagon's inspector general started taking an interest in the DIVAD program. The GAO also questioned the Army's assessment of DIVAD.

By January 1984, the Army began testing the first production model of DIVAD, now named after famed sharpshooter Sergeant Alvin York, a World War I Medal of Honor winner. The Army said that the tests were inconclusive, but sent a letter to Ford the next month complaining that work on the DIVAD did not measure up. Undersecretary of Defense for research and engineering Richard DeLauer then told the Army to do more realistic testing before any more DIVADs were purchased. That July, the Army tested three production-model Sergeant Yorks, again claiming the tests were inconclusive.

Still more testing was done in the fall of 1984 at Fort Ord by the Army's Test and Evaluation Agency. The agency noted that the Sergeant York's radar system had problems tracking target aircraft, but that these problems could be fixed. Still, a 100-page report by the same agency noted that the Sergeant York did not deliver much improvement over the Vulcan and Chaparral systems it was supposed to replace. "In terms of its ability to acquire targets, track them, and engage a mix of targets and survive in the field, the DIVAD comes up a loser in almost every scenario," the report said.

Bad news about the Sergeant York was becoming commonplace in the newspapers. Major General C.D. Bussey, chief public affairs officer for the Department of the Army, took the *New York Times* to task for what he described as a negative and unfair editorial criticizing the DIVAD program. The Sergeant York had to be radar-aimed to hit high-performance aircraft, and didn't have to hit aircraft maneuvering at greater than two Gs since nothing can do a bomb run at that level of evasion, he explained. The thin-skinned helicopters of the Vietnam War could be shot down with ease, but the Soviets were fielding armored gunships that could only be shot down by the DIVAD, he added.

"We have not done a good enough job of explaining the DIVAD program," Bussey admitted. "[W]e would welcome the opportunity to do that . . . . As a minimum, perhaps we can stop recycling myths and misconceptions. Your editorial is a case in point."

Then in April 1985, things went from bad to worse.

The Sergeant York program landed on the desk of Secretary of Defense Caspar Weinberger.

Weinberger was nicknamed "Cap the Knife" back when he was Secretary for Health, Education and Welfare in the Nixon Administration. While he used to cut budgets at HEW, he rarely touched the budget scalpel, much less the budget ax or budget chainsaw, while squiring weapons programs at Defense.

It seemed that Weinberger never met a weapons system he

didn't like, as the Pentagon budget shot up from $200 billion to $300 billion on his watch. In the rush to rearm, many weapons systems were being purchased with ongoing problems and vexing flaws. Production models of the F-18 Hornet were plagued by tail cracks. The B-1 bomber never seemed to work as advertised. The stuff that made it to production was much pricier than predecessor systems. Critics began to question the value of defense purchases as higher spending without tax increases resulted in higher deficits. Tales of $600 toilet seats and $1,200 coffee pots plagued the debate on defense spending.

Sergeant York was caught in the spotlight, becoming the cheap poster child for much that went wrong during the $1 trillion, five-year Reagan defense program. In the midst of this debate, the mess ascended to Weinberger's desk, where it received scrutiny with a magnifying glass rather than a rubber-stamp approval. Weinberger now had a cheap opportunity to get tough on wasteful defense spending, and Sergeant York was going to be the whipping boy.

Another round of more realistic testing was slated for the summer of 1985.

If the DIVAD performed as advertised, it would be kept.

If it failed, it would be cut.

The Army had one more chance to get it right.

So the gun got its final test and the Pentagon's Office of Operational Testing and Evaluation got to watch. "As tested, the Sergeant York was not operationally effective in adequately protecting the friendly force during simulated combat." The self-propelled AA gun kept breaking down. The Army considered whether to supplement the twin Bofors guns with Stinger missiles to help it shoot down enemy aircraft, which would have made moot the reason of having the guns in the first place.

Failure on the test range was compounded by obsolescence.

When the DIVAD was conceived in the late 1970s, it was expected to engage enemy aircraft at ranges up to 2.5 miles, better than the 750 yards or so that the Vulcan could deliver. By the early 1980s, the Pentagon was getting field reports about the Soviet Mi-24's debut in Afghanistan, where this armored helicopter gunship was firing its missiles at Afghan mujahedeen at ranges up to 4.6 miles. When the Mi-24 was simulated during DIVAD's final test, there was no way a gun with a 2.5-mile range was going to hit a helicopter sniping at it from three miles away.

In August 1985, Weinberger took the knife to the Sergeant York, killing the program with one stroke.

The Sergeant York failed after consuming over $1.5 billion in development money. More than sixty DIVADs were built at $8 million each by the time Weinberger terminated the program. The Army needed to find a replacement for its replacement. So it planned to spend $8 billion on a collection of mobile Stinger missile launchers, a new radio/computer targeting integration system, a fiber-optic guided missile that could hit airborne targets six miles away, and a retrofit of existing vehicles to possess some anti-aircraft capability.

Meanwhile, the Soviet Mi-24 gunship, which eliminated the need for the Sergeant York, was in turn defeated by the high-tech Stinger shoulder-fired heat-seeking missile, handled by often illiterate Afghan tribesman fighting the Russian invaders.

Who killed Sergeant York?

Weinberger was only guilty of mercy killing when he put the DIVAD out of its misery. Blame first the Pentagon for thinking it could get a new weapons system on the cheap. Blame second the contractors who promised more than they could deliver—at a fixed price.

The DIVAD contract drew five competitive offers. The program was supposed to use off-the-shelf components of proven reliability, delivering an AA gun with warranties, a fixed price, no razzle-dazzle, no cost overruns, and no delays. This technique had worked before, which is how the Army got its enormously successful Multiple Launched Rocket System (MLRS).

The Army's "hands off" approach complemented the joint failure of Ford Aerospace and General Dynamics to "get it right." Had the Army been closer to the program, problems could have been picked up earlier and fixed. Even so, the Army always asked the Pentagon for another chance to fix DIVAD's problems, again and again and again.

Representative Dennis Smith (R-Ore.), a persistent critic of the Sergeant York, blamed the Pentagon. No complex weapons system should be purchased unless it could be realistically tested first, he said.

James R. Ambrose, undersecretary of the Army, disagreed. Technology is America's strong suit in weapons development, and testing should be continuous to allow for a new weapon to be developed as it is being deployed. Experience in the field would beget more improvement, he explained. "Most of the investment in the military goes without such testing," as was done with DIVAD, Ambrose said. "You do it with some use of analytic models, play computer games, and a lot of it is military judgment."

Even the Department of Justice got to place some blame on General Dynamics, indicting four of its executives involved with overseeing the DIVAD program in December 1985. One of those tagged was James Beggs, who was heading NASA when the indictment was served.

The substance of the case was small potatoes. General Dynamics was accused of shifting a $3.3 million cost overrun on a

$40 million fixed cost DIVAD contract to two research accounts that the government reimbursed. (This at a time when the U.S. government was accused of wasting tens of billions of dollars on shoddy weapons.) The case so lacked substance that U.S. district court judge Ferdinand Fernandez dismissed it in June 1987 on the recommendation of U.S. attorney Stephen Czuleger, the prosecutor handling the case.

"I'm going to ask for an apology," said Beggs, who had to resign from NASA to prepare for his defense. "I certainly have no objection to the defense industry being put under a magnifying glass. What I do object to is this rush to justice in bringing what appeared to be very bad cases."

But the last word in the blame game went to William Weld, then assistant attorney general heading the Justice Department's criminal division. In June 1987, his unit found that defense contractors were allowed by the Pentagon to understate their development costs at the taxpayer's expense. This undercut the ability of federal prosecutors to go after defense contractors who abused contract terms. "The military officers may overlook or ignore the inactions by defense contractors, not because of evil intent, but because of a belief that the importance of the project or new technology has to national security," Weld said.

The contractors cheated to maintain contract goals and turn a profit.

The military wanted to accomplish the weapons development mission.

Everybody does it.

No one is guilty.

And the taxpayer picks up the tab.

> "If you believe the doctors, nothing is wholesome; if you believe the theologians, nothing is innocent; if you believe the military, nothing is safe."
>
> —Lord Salisbury

The Expensive Pipe Dream of Missile Defense

Douglas Niles and Donald Niles, Sr.

If the Enemy Puts Tracking Beacons in Their Rockets It *Might* Work . . .

"Star Wars": The very name evokes high-tech glamour, scientific accomplishment bordering on the magical, and a certain triumph of the forces of light over darkness. The term has become—even without the permission of George Lucas—descriptive of the United States Missile Defense program, more formally known as the Strategic Defense Initiative. Originally brought to the public's attention by President Reagan during the 1980s, there is some possibility that the idea of the program (the system itself was decades away from implementation) contributed to the downfall of the Soviet Union simply because it symbolized American capability and ambitions in a way that highlighted the decaying dream of the U.S.S.R.

Of course, strategic nuclear weapons represent the single greatest threat to human life with which humankind has ever menaced itself. During the decades of the Cold War, the United States and the Soviet Union virtually bristled with lethal, long range, unstoppable missiles—weapons with which, as has been well documented, each power could have completely obliterated the other, probably several times over. The reason they didn't was a policy that, in theory, sounds mad, and in actuality was called MAD: Mutually Assured Destruction.

The theory assumed that neither side in this incredibly high-stakes gamble would dare to take the first shot, because they could not destroy the other side's ability to retaliate in such force that the nation who started the war would also be completely destroyed. Thus, though in reality each side could obliterate the other, neither side was willing to initiate a nuclear war.

Those days are gone now, of course. The United States, for the moment, is the only global superpower. And even though America has removed some of its missiles from service, others remain ready to be launched at a moment's notice. Furthermore, during the first decade of the twenty-first century, the government has pressed forward with a new generation of highly expensive, terribly lethal nuclear weapons.

At the same time, the country has spent tens of billions of dollars—more money than on any other weapons system—for a vast and complicated array of equipment that might not even work, and is not designed to face any threat that could conceivably be directed our way.

Over the early years of the twenty-first century, the MD system was tested multiple times, and yielded an uninterrupted string of failures. Intercepting missiles failed to launch, or to separate from booster rockets. Those that did fly successfully failed to intercept the incoming missiles. After most of these flops, the Pentagon labeled the test a "limited success" because even an utter and complete failure potentially offers an opportunity to gain some useful data. On other occasions, such as an abject failure in May 2007, when the target missile actually failed to get high enough to activate the system, the Pentagon alters the terminology to describe the failure as a "non-test."

Even to attain the level of failure that has been a persistent feature of this program, the government has gone to ridiculous and unrealistic lengths to make it easier for the intercepting missile to find the target rocket. One way that the tests were modified was to enhance the target missiles with tracking beacons to make them easier to locate and intercept—and even then, the intercepting missile could not hit the incoming rocket. Nevertheless, perhaps diplomatic efforts will be next, in an effort to persuade potential enemies to put those tracking beacons in their own missiles.

Of course, that would only be possible if there was an enemy out there that might, possibly, make this system worthwhile. Right now, however, that doesn't seem to be the case.

It is indicative of the tremendous waste of money and resources invested in the MD system that the threat we are hoping to defend against does not actually exist. This is no longer Reagan's "Star Wars" idea. There is no practical thought that, in the event of a catastrophic nuclear war, the MD system would be able to take out every one of the hundreds of warheads that might conceivably be sent our way by another superpower. Instead, the missile defense system has been pushed through

Congress with the argument that it might potentially protect the country from a single missile launched by a rogue state—North Korea being the most commonly mentioned potential adversary.

While no one would argue that the North Korean government is peaceful, stable, or even moderately rational, this argument overlooks a key fact: if they, or some other country, wished to deliver a nuclear bomb against an American target, they wouldn't do it with a ballistic missile. For one thing, the launching point of such a missile would be immediately and accurately placed by satellite and other reconnaissance technology. Any country launching a single nuclear missile against the United States can be virtually assured of its immediate and complete annihilation by an overwhelming response.

It is far more likely that such a bomb would be delivered by secretive means, smuggled in by ship, for example, or flown aboard a small plane, or even brought across the border by truck or van. The MD system would protect the country against none of these possibilities, though it must be admitted that these are very real, and terrifying, threats.

Instead, the government has publicly declared—in a pointed message to North Korea—that the untested, non-functioning missile defense system has actually been "deployed," with the stationing of some potential interceptors in Alaska and elsewhere. Still, there is virtually no chance it would work if actually challenged. Assuming the interceptor could be launched on time, it would take a lucky shot, akin to using a gun to shoot another gun's bullet out of the air, for an incoming missile to be destroyed.

Among the many tenets of warfare being ignored, at tens of billions of dollars' cost, is that an enemy that knows about a defensive system is almost certain to take steps to defeat that system. If, by some miracle of technology and spending, the United States eventually comes up with a means of detecting an enemy missile launch, launching an intercepting missile in the few minutes available, and blowing the incoming missile out of the sky, the planners and backers of the MD program seem to think that an enemy will not change his tactics to suit the defense.

However, there are lots of ways—all of them much simpler than creating an intercepting missile system—for another country to make it impossible for the defense to work. Missiles could be disguised or masked with stealth technology, or equipped with electronic interference devices that could negate the MD tracking system. The most basic, and inexpensive, idea would be for an attacker to launch a whole host of decoy missiles, with only one or two rockets actually armed with nuclear warheads. The intercepting missile system would have no way of distinguishing the real targets from the decoys, and could simply be overwhelmed by a number of fakes. If the United States then goes ahead and builds enough—very, very expensive—MD batteries to take out a lot of decoys, the attacker can simply build more cheap decoys. And still, the missile defense system wouldn't know which missiles to intercept and which to let through.

So, do *you* feel lucky?

BOOKS BY
BILL FAWCETT

YOU SAID WHAT?
Lies and Propaganda Throughout History
ISBN 978-0-06-113050-2 (paperback)

From the dawn of man to the War on
Terror, Fawcett chronicles the vast history
of frauds, deceptions, propaganda, and
trickery from governments, corporations,
historians, and everyone in between.

OVAL OFFICE ODDITIES
**An Irreverent Collection of Presidential
Facts, Follies, and Foibles**
ISBN 978-0-06-134617-0 (paperback)

Featuring hundreds of strange and
wonderful facts about past American
presidents, first ladies, and veeps, readers
will learn all about presidential gaffes, love
lives, and odd habits.

HOW TO LOSE A BATTLE
Foolish Plans and Great Military Blunders
ISBN 978-0-06-076024-3 (paperback)

Whether a result of lack of planning,
miscalculation, a leader's ego, or spy
infiltration, this compendium chronicles
the worst military defeats and looks at
what caused each battlefield blunder.

YOU DID WHAT?
Mad Plans and Great Historical Disasters
ISBN 978-0-06-053250-5 (paperback)

History has never been more fun than it is
in this fact-filled compendium of historical
catastrophes and embarrassingly bad ideas.

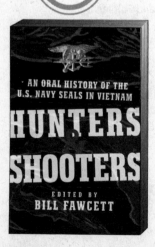

HUNTERS & SHOOTERS
An Oral History of the U.S. Navy SEALs in Vietnam
ISBN 978-0-06-137566-8 (paperback)

Fifteen former SEALs share their vivid,
first-person remembrances of action in
Vietnam—brutal, honest, and thrilling
stories revealing astonishing truths that
will only add strength to the SEAL
legacy.